BEYOND CLONING

Science has suddenly presented us with another awesome challenge. Biologists have long known that DNA —minute strands of protein in every living cell—were the building blocks of life. Only now they have discovered exotic new ways to graft DNA from one cell onto the genes of a completely different organism. A tiny slip-up could have staggering consequences. Future generations could curse or bless us for what we do now.

Michael Rogers weaves together the electrifying story of the new research, the drama and clash of personalities at the strange assemblage of brilliant scientists that met in a secluded retreat on the Big Sur, California coast to regulate themselves, and the intricate moral questions involved in a worldwide controversy that faces us all.

"Entertaining . . . marvelously lucid." *Newsweek*

"Well-researched, competent, and accurate . . . consistently excellent." *New Haven Register*

"With a vivid writing style . . . Rogers presents the science colorfully yet accurately." *Science Magazine*

"Rogers commands an informal penetrating style rare among science writers." *Saturday Review*

BIOHAZARD

MICHAEL ROGERS

AVON
PUBLISHERS OF BARD, CAMELOT AND DISCUS BOOKS

AVON BOOKS
A division of
The Hearst Corporation
959 Eighth Avenue
New York, New York 10019

Copyright © 1973, 1976, 1977 by Michael Rogers
Published by arrangement with Alfred A. Knopf, Inc.
Library of Congress Catalog Card Number: 77-74986
ISBN: 0-380-41731-6

First Avon Printing, March, 1979

AVON TRADEMARK REG. U.S. PAT. OFF. AND IN
OTHER COUNTRIES, MARCA REGISTRADA, HECHO EN U.S.A.

Printed in the U.S.A.

For Jan, with all my love

There is a name that is commonly applied to such a competition for a single source of raw materials among a number of self-aggrandizing chemical systems, each of which is subject to small random changes affecting its rate of activity—it is called *evolution*.

—Dean Wooldridge,
Mechanical Man

If progress is a myth, that is to say, if faced by the work involved, we can say: "What's the good of it all?" our efforts will flag. With that the whole of evolution will come to a halt—because we are evolution.

—Teilhard de Chardin,
The Phenomenon of Man

Contents

Acknowledgments

The layman who writes a book about science ends up owing much to the patience of many people. One who writes about a complex new area of science and its relationship to society is bound to incur even greater debts. What follows is only a partial list of the many people who have been kind with their time and attention: Emmet Barkeley, Albert Bothwell, A. M. Chakrabarty, Stanley Cohen, William J. Gartland, Richard Goldstein, Rae Goodell, Howard Lewis, Norman Metzger, Colin Norman, Tabitha Powledge, Maxine Singer, John Tooze, James Watson, Charles Weiner, and Susan Wright.

Jann Wenner gave generous space in *Rolling Stone* to my initial curiosity about recombinant DNA at a time when the subject was thoroughly arcane.

Paul Berg displayed both patience and endurance in reading this manuscript and offering technical advice and corrections—although he should not be held responsible for any errors or misinterpretations that may have escaped his notice.

And my wife, Janet Hopson, provided me early tutoring in basic biology, careful reading at every stage of the manuscript, and the constant moral support without which this project could never have been completed.

Michael Rogers
March 17, 1977

1

The Beginning

One afternoon late in February 1975, I was unpacking my suitcase on a narrow bed in a redwood dormitory, on the grounds of a California conference center called Asilomar. Three hours to the north was San Francisco; three minutes' walk to the west was the Pacific, framed by Monterey pine and rock-bound tidepools. The weather was transparently clear, almost unnaturally crisp—the kind of perfect climate the Monterey peninsula offers regularly to midwinter refugees from less beneficent climes.

My own suitcase this visit, however, was all business: half filled with clothes and half packed with textbooks on biology, microbiology, and molecular biology. My visit to the Monterey peninsula was for neither tennis nor golf, but rather to attend a moderately obscure event called the International Conference on Recombinant DNA Molecules.

Not until four days later did anyone think of taking a group portrait at Asilomar, and by then the official photographer had already departed. And thus that diverse mix of 140 scientists, who were manipulating the most fundamental of life processes in laboratories from Moscow to Memphis, remains pictorially unrepresented in the history of modern science.

But their activities will not go unrecorded: The conference—four intense twelve-hour days of deliberation on the ethics of genetic manipulation—provides too neat and apt an epoch marker for any historian to ignore. And while I was hardly a historian, recording this event was exactly what I was doing at Asilomar that warm afternoon. My role, essentially, was that of spectator: On the basis of a few brief notes in technical journals, I'd come to press my nose against the glass of frontier technology and see, if I was lucky, history in the making. This unprecedented process of scientific self-regulation would, I suspected, positively glow with the aura of history. And the science, I blithely assumed—armed as I was with that suitcase of textbooks—I would pick up as I went along.

The significance of the Asilomar Conference was even then tangible and had been, in fact, since the previous summer, when a terse letter signed by eleven leading life scientists and headed "Potential Biohazards of Recombinant DNA Molecules" had appeared simultaneously in three major science journals in the United States and Great Britain.

The letter had bluntly called for a universal moratorium on certain experiments with a brand-new technique called recombinant DNA engineering until their implications for public health and planetary ecology could be further considered. It was the first such appeal in history to be made to the global scientific community. And it dealt, moreover, with a particularly fundamental kind of research. Over a matter of months, the young science of molecular genetics had happened upon its most powerful tool to date: the ability to manufacture, in the laboratory, creatures never before seen on the planet and, in all likelihood, impossible before human intervention. "Nature," as one microbiologist noted at Asilomar, "does not need to be legislated. But playing God does."

The phrase was dramatic, but then so was the research itself. It had only been a century earlier that the Austrian monk Gregor Mendel, toiling in his monastery pea-patch, first described the apparent logic of genetic inheritance. And it was only far more recently that human beings even identified the minute chemical medium in which that genetic information is stored and transmitted.

The medium is, of course, deoxyribonucleic acid—DNA—an intricate, lengthy. twisting organic molecule that carries the design of every living creature on the planet, from the paramecium in the mud puddle to Albert Einstein, in one variation or another of its sinuous molecules. A single strand of human DNA, microscopically small, contains the information of at least one thousand thick books. The chemical keys to that library took a century to discover. To translate one volume appeared even more difficult; and to write one's own book —impossible.

Or perhaps not. "Science," as a British biologist observed at Asilomar, "has built-in pauses; some last a hundred years. But the thing about recombinant DNA is that it's suddenly made many things very easy that were once very difficult." Recombinant DNA, in a sense, represented the discovery of the first rudiments of grammar for that previously unspeakable genetic tongue.

But what was recombinant DNA? Greek mythology describes a creature called a chimera, a female monster composed of pieces of two or more animals. In the months before the Asilomar conference, molecular geneticists had begun to create chimeric DNA molecules. Recombinant DNA techniques used certain newly discovered chemicals to disassemble the long DNA molecule in so orderly a fashion that the loose bits of genetic information could then be recombined, chemically, into a coherent sentence. And such a recombinant sentence may well describe—and function like—the genetic blend of two altogether different creatures, incapable of mating in nature, like "joining duck DNA," as one of the early English workers was fond of saying, "with orange DNA."

In early 1975, however, the new techniques hardly aspired to either duck or orange DNA; they worked, essentially, only with bacteria and viruses—organisms so small that most human beings really only notice them when they make us ill.

But that, right there, was the problem. Although recombinant DNA techniques made it possible to perform a new set of novel and important experiments, they also allowed the transfer of a whole range of potentially harmful characteristics, from antibiotic resistance to cancer production, into microorganisms that did not intially possess those traits—creating, at worst, novel diseases for which no cure might exist. The initial fear, moreover, was not that someone might do so on purpose, but rather that novel microorganisms would be created and released altogether accidentally, in the innocent course of legitimate research. "Such experiments," the moratorium letter had warned, "should not be undertaken lightly." But that was, in fact, part of the problem: Nobody in the field was taking them lightly at all. Recombinant DNA was so potent a research tool that if left to itself in the competitive arena of experimental science, it would almost certainly soon grow entirely uncontrollable.

Hence the moratorium letter. And hence, nine months later, the meeting at Asilomar. The International Conference on Recombinant DNA Molecules was something far different from the average scientific meeting, and from the outset, it marked a watershed in the manner in which scientists approached their ability to control nature. It seemed more than coincidence that this first instance of voluntary scientific self-regulation also involved the first real techniques of manipulating DNA itself—our initial foray into what François Jacob once called "the ultimate biological invariable."

Asilomar marked the beginning of a period of scientific self-searching that may ultimately prove equal to that faced by nuclear physicists in the years prior to sustained fission. Only instead of occurring in the depths of wartime secrecy, in a made-over squash court beneath a Chicago stadium, this time the proceedings all took place in public.

After twenty minutes of the Monday morning session at Asilomar, I suspected I was right about watching history—and utterly wrong about picking up the science. Eight hours later I would be back in my dormitory room ransacking my suitcase of textbooks. The next day someone would gently inform me that some of the words might not even be in my borrowed textbooks, as the language itself was quite simply too new. With that, I settled back to watch three more days of history-in-the-making on the California coast from the distinctly uncomfortable vantage point of almost total technical ignorance.

One of the many popular treatments about the future of genetic engineering called DNA "a once-esoteric set of initials now just about as familiar to the American household as PTA or DDT." The initials, I realized at Asilomar, may be familiar, but the concepts behind them remain far less accessible. And I discovered, in the year to come, that while molecular genetics will almost certainly have as much impact on the next two decades as nuclear physics has had on the previous two, the quickest way to generate epidemic boredom in polite company is to use, too often, words like DNA or chromosome.

At Asilomar, I'd assumed the fault was my own. My post-Sputnik education had been anchored, firmly, in the physical sciences, and while I'd strayed far from those disciplines, I nonetheless still shared the archaic hard-science prejudice that biology is an inherently soft undertaking involving small furry animals that display no dependably replicable phenomena apart from regular excretion.

How archaic my prejudice was I learned early on: The whole field of molecular genetics was really born with a mass influx of renegade physicists in the years surrounding World War II. As early as 1932, in fact, Niels Bohr, the elder statesman of quantum mechanics, had proffered the notion—a curiously backhanded sort of vitalism—that the laws of physics might not, in all, be sufficient to explain the phenomenon of life; that some

17

uncertainty principle, similar to that encountered on the subatomic level, might function as well in biology. Immediately following World War II, another premier physicist, Erwin Schrödinger, published a thin volume called *What Is Life?*, which took Bohr's notion one step further by suggesting that there might be yet unknown principles that could only be uncovered by investigating the physical basis of life. For post-Bomb physicists, Schrödinger's suggestion of fresh pastures launched something of an interdisciplinary land rush—and by the early 1950s, molecular genetics had been launched with a whole new complement of strayed physicists, a development that would change the shape of biology to come just as certainly as Copernicus changed the direction of astronomy.

Yet the leading edge of that science—the kind of biology that physicists entered the field to do—seems to remain, for the layman, less accessible than even the workings of a standard nuclear power plant. But then nuclear fission—at least since that December day in wartime Chicago—has been somewhat under the influence of human volition, while the replication and expression of DNA molecules, regardless of how intrinsically interesting, has remained beyond that same kind of human control. That is, until recently. And yet I found myself at Asilomar, listening to arguments about the experimental manipulation of DNA in terms that, to my untutored ear, bordered on science fiction.

The whole matter of "genetic engineering" has, of course, received most of its popular attention in science fiction—ever since that day in 1931 when Aldous Huxley sat down in his study by the Mediterranean to compose an answer to H. G. Wells's optimistic views on the effects of applied science—a book that Huxley finally named *Brave New World*.

The effort didn't please Wells—who promptly labeled it "treason to science and defeatist pessimism"—but Huxley's notions of decanted clones have colored even the most serious treatments of "genetic engineering" for decades. And by now, a small army of academic thinkers has attacked the subject from various social and ethical directions and left it in thorough disarray. An apt summation of the argument to date might come from the postscript of one recent impassioned book on the subject by a Columbia University sociologist: "Ideas are common," he wrote of the continuing debate over genetic engineering, but "effective reforms are rare and require long, sustained labor."

The sentiment applies neatly to the process represented by Asilomar. Ideas have been common, and have usually been woven around complex visions of dial-a-baby and precisely how parents will choose between blue eyes or brown eyes, feet, flippers, or wings. Yet throughout the course of those colorful musings, no one even briefly suspected that the first time human genetic meddling would grow truly dangerous—and able to alter life as we know it—it would be at Asilomar, in an arena populated by creatures visible only by microscope.

All of which leads one to wonder exactly what the phrase "genetic engineering" means in the first place. In the broadest sense, genetic engineering is a practical business human beings pursued long before science gave it a name: the calculated breeding of plants and animals to reorganize, through mating, the physical characteristics of related organisms.

A useful distinction here was introduced in 1909, by a biologist named Wilhelm Johannsen. Johannsen suggested two words that nestled promptly into the vocabulary of modern biology: *genotype* and *phenotype*. Genotype refers to the genetic information—the "genome," in the form of DNA molecules—within a given organism. Phenotype refers to the manner in which that information—those genes—read out in the real world: the shape and constitution of the organism itself.

Before recombinant DNA, the only ways that human beings could alter genomes were essentially nature's ways. There was selective cross-breeding, which requires

19

not only nature's own time—often many generations—but also, even in the case of artificial insemination, some aping of nature's methods.

More recently, there has been the far more technological process of mutation—exposing a given genome to radiation or chemicals and then waiting to see what that disruption will produce in the next generation. Such deliberate mutation has given us, for example, the seed packets advertised in the backs of magazines and guaranteed to yield, if nothing else, *curiosities*. It has also, as we will see, proven an exceedingly valuable research tool. But mutagenesis is, nonetheless, a technique that has been inflicted by cosmic rays since long before humans were even aware of a cosmos, and which is, in the last analysis, the genetic equivalent of pinball.

In a sense, both procedures could fairly be called genetic engineering. Recombinant DNA, however, is somewhat different. The new techniques allow one to circumvent chemically the natural processes that over millennia have sorted and polished individual genomes —to pluck out and reinsert individual segments of genetic information chemically, in purest molecular form, and to do so with at least some degree of foresight and choice. Crude as the new technique was in 1975, it nonetheless seemed to imply a new and unprecedented human control of evolution—a final expulsion from the Garden, an abrupt initiation into what the phrase "genetic engineering" might really imply, and thus a chilling sort of new responsibility.

That, at least, is what turned my brief visit to Asilomar into a year-long obsession. And while the whole notion seemed a bit arcane at the outset, those four days in California ultimately captured a remarkable amount of attention in the months that followed. Newspaper coverage was extensive; national magazines from *Reader's Digest* to *Rolling Stone* devoted space to the matter; college undergraduates dissected the subject in honors theses; seminars, symposia, and classes arose from Berkeley and Stanford to Harvard and Yale; a television special went into production; MIT even established an elaborate archive documenting the oral history of the event.

And even after a year of reading, I find that the phrase "molecular genetics" still holds an odd thrill for me. It represents the fusion of two exceedingly powerful concepts—the notion of heredity, and the notion that beneath the surface of life, mechanisms lie in discrete units of inanimate matter.

It all conjures up a world of unabashed mechanism—a step beyond that opening quote by Dean Woodridge—an analysis of life processes that, as I listened, increasingly embraced the language of engineers. And at the same time, it seemed a world where the ordinary processes of evolution were about to turn around, on a practical level that de Chardin probably little suspected when he wrote that we *are* evolution.

And all this arose from a science only embryonic during that optimistic Jesuit's recent lifetime. Yet from the first day at Asilomar, it was obvious that the vocabulary of recombinant DNA had already grown as abstruse and fluid as that of particle physics. By then I had already decided that I could survive to ripe old age and never really suffer from my ignorance of the difference between mu- and pi-mesons. But someday, the precise manipulation of the material between the sugar-phosphate backbones of a DNA molecule will almost certainly, in some way I likely cannot even imagine, affect all my children.

My curiosity has by now led me into an odd world of bizarre microorganisms, exotic biochemicals, gene therapy, patent law, drug manufacture, official biohazard symbols, designs for high-containment greenhouses, and science fiction phrases like "naked DNA" or "gene transfer in the gut." But in the weeks following Asilomar, my questions were more basic, and they began, in fact, with a simple one:

What *is* this DNA?

2

A Collision
of Disciplines

The answer to the question "What is DNA?" involves a story—a story about the collision of two separate lines of scientific inquiry that fused, in the end, into a single and exceedingly powerful discipline that finally produced the controversy at Asilomar.

That collision took place only a few decades ago, although its component parts began their acceleration a century earlier—in the late 1860s, in Tübingen, Germany, where a new M.D. named Friedrich Miescher started buying used bandages at bargain rates from the local surgical clinic. Miescher promptly discarded the bandages but kept the pus found therein—which he removed, by careful laundering, and then subjected to complex chemical analyses.

The chemistry of living creatures was just then becoming a matter of scientific interest, and Miescher's pus did not disappoint him. He found a previously undefined assemblage of molecules that he called "nuclein." The odd material occurred frequently enough in nature that by 1871, Miescher had isolated it as well in egg yolk, and other researchers noted nuclein in cells from birds, snakes, and yeast cultures. "There is a group of nucleins," Miescher concluded cautiously, "and it will surely increase with additional members."

Miescher moved to the Swiss city of Basel, on the Rhine, and proceeded to find nuclein in the rich spermatozoa of that river's spawning salmon. But soon—when support from his contemporaries seemed less than enthusiastic—he changed his focus from the chemistry of Rhine salmon sperm to the somewhat more tangible physiology of the salmon itself.

By 1900 or so, however, Miescher's odd molecule had been well accepted: It was indeed a curious cellular component and, moreover, had earned the more complex designation of "nucleic acid." Within fifty years it would be recognized as two separate compounds dubbed, rather chummily, DNA and RNA. A great deal, however would come before that.

The second line of scientific inquiry—the other half of the collision—also began in 1900, when three European botanists independently rediscovered a scientific paper that had moldered in dozens of libraries for nearly thirty-five years. The paper was called "Experiments on Plant Hybrids," and dealt with the matter of inheritance among sweet pea plants, as observed by an obscure Austrian monk named Gregor Mendel.

According to Mendel, inherited characteristics among pea plants—qualities like the shape or color of seeds—are transmitted from parent to offspring by a very definite, mathematical pattern of succession. And the reason that the three turn-of-the-century botanists had rediscovered Mendel's long-lost offering was that they had all, independently, noticed exactly the same thing. There was in fact a very precise calculus of heredity.

Now the notion of inherited characteristics was by itself not new. It had been observed by Hippocrates around 400 B.C., and in the centuries following it had served as the basis for everything from animal husbandry to adultery proceedings. What Mendel and his rediscoverers were suggesting was another matter altogether: that heredity operates on some fairly mathematical, occasionally calculable basis. And that theory implied, moreover, individual units of heredity that could affect the generation of offspring on a mathematically predictable basis.

The notion of discrete units of heredity had in fact existed for some time, spawned by the microscopic study of egg fertilization. Something, clearly, was transmitting information between sperm and egg, and even before the process by which it did so was understood, researchers set out to name the unknown: "physiologic units," Herbert Spencer suggested in 1867; "gemmules," offered Darwin in 1869; "plastidules," said Elsberg and Haeckel; "stirps," proposed Francis Galton. By 1909 the names had run from "pangens" and "bioblasts" through "biophors," "plasomes," and "idioblasts," and had finally settled on Wilhelm Johannsen's altogether anticlimactic "gene."

Johannsen, clearly, had a way with a name, and as noted earlier, coined along with "gene" the designations "genotype" and "phenotype." That apt distinction made it incumbent upon anyone who really wanted to pursue the mysterious workings of the newly dubbed genes to come up with some hard connection between those theoretical pieces of genetic machinery and the characteristics observed in the creature itself. The mathematics of Mendel and his followers provided highly suggestive evidence for that relationship. The question now was how to verify that connection experimentally. And what that called for, clearly, was some way to see (or even handle) the genetic material itself—the process which, quite naturally, finally yielded the technique of artificial DNA recombination.

That process has been played out by what must be one of the most disparate experimental casts in modern science: It begins with a fruit fly, continues with a bright-orange bread mold, accelerates with a minute bacteria whose relatives reside in the bowels of every individual holding this book, and culminates in a predatory virus so small that its only victims are germs.

In 1910, Thomas Hunt Morgan chose the tiny fruit fly *Drosophila* to be his probe into the mechanics of genetics. *Drosophila* are minute creatures who reproduce many times over in the time a single crop of Gregor Mendel's sweet peas took to bloom. Not to mention that Morgan's fruit flies were content with small glass bottles and mashed fruit rather than monastery gardens.

Morgan began, however, much as Mendel did—by identifying specific physical characteristics in his organisms that he could track through successive generations. For Mendel, it had been flowers and pods; for Morgan, it was more wing size and eye color.

But while the characteristics Mendel chose to study were fairly common pea plant traits, Morgan's markers were true mutants—far rarer physical modifications due to the mutation of a single gene, characteristics as rare in fruit flies as, say, albinism is in human beings. Morgan and his students spent five years examining hundreds of thousands of the little creatures before identifying enough mutants to carry on the experiment. (One of Morgan's students, H. J. Muller, discovered ten years later that by exposing his fruit flies to X-rays, the rate of mutation could be increased greatly—thus reducing drastically the total number of *Drosophila* he had to examine.)

Morgan, at the outset, confirmed Mendel's description of the mathematics of heredity. But he went one step further.

Chromosomes—the dark, rodlike objects observed within the nucleus of all animal cells—had long been suspected of being responsible for at least some portion of the hereditary process. The chromosomes seemed unusually active, in fact, during the process of reproduction, exhibiting in particular a peculiar sort of microscopic mating dance called "crossing over."

"Crossing over" referred to the way two chromosomes in a fertilized germ cell, on its way to becoming a fruit fly larva, would come together and then separate—in the process trading some of their genetic information. In the case of his mutant fruit flies, Morgan watched that phenomenon carefully, and then noted exactly which mutations expressed themselves in terms of the finished fruit fly. With those two pieces of information, Morgan reasoned, one could ultimately figure where each mutant gene actually resided on a given chromosome. A fruit fly has eight chromosomes (compared to forty-six for human beings), and by 1926 Morgan had published a historic document: a map of several fruit fly chromosomes, indicating, on the basis of his crossing-over experiments, the location of each gene controlling certain functions—wing size, eye color, etc. It represented, quite literally, the first mapping of unknown genetic territory—a process, as we will see, that continues with vastly more sophisticated technology today.

Mapping made these tiny bits of genetic matter seem far more real. And in the next few years, those maps grew in complexity. But the actual knowledge of what genes did changed very little. How does a gene—whatever it is—control the shape of a man?

One hint had already arisen, more than a decade before Morgan published his first maps, when Archibald E. Garrod—a man distinctly ahead of his time—coined the phrase "inborn errors of metabolism" in referring to a hereditary type of arthritis apparently induced by the failure of the body to break down a certain waste component carried in the blood. Garrod knew that the process of breaking down a substance into simpler substances —metabolism—is encouraged by an enzyme, a class of compounds that aids the completion of chemical reactions. Since the arthritic condition was caused by a missing or defective gene, he reasoned that the gene in question might in fact function by ordering up the production of the required enzyme. At the time, Garrod's observation slid rather quietly by, even though it was a striking precognition of a theory that, decades later, would be called precisely "one gene—one enzyme."

That first major identification of genes with a specific chemical function waited, however, for nearly thirty years, for a pair of researchers named George Beadle and Edward Tatum, and a second character in the experimental laboratory, a bright orange bread mold called *Neurospora crassa*. From sweet peas to fruit flies to bread mold: Already the pattern of moving toward increasingly simple biological systems was clear, and inherent in that was a remarkable assumption: that the genetic functioning of living creatures is sufficiently similar all across the kingdom of life that one's choice of subject can be based, in large part, on experimental convenience.

Beadle and Tatum set out to grow enough bread mold to discover, as Mendel and Morgan before them, mutants that could be readily identified as separate from the rest of the sample. At this level of life, however, one had neither pods nor wings to consider, so instead, Beadle and Tatum isolated *Neurospora* mutants distinguished by their inability to grow unless provided with a particular nutrient that, by contrast, a normal healthy *Neurospora* could make for itself out of its standard diet of sugar and salts. On the same diet, however, the mutant bread mold would die, and only when the specific nutrient that the bread mold was unable to make for itself was added could it survive.

The concept of using a mutant microorganism unable to survive without a specific "growth factor" remains as potent a tool in modern molecular genetics as it was for Beadle and Tatum: After the mutant bread mold had been crossbred at length, they managed to prove that its inability to synthesize all the vital nutrients from its standard diet was a direct malfunction of a single gene. And then, with a bit of biochemistry, it was a short step to realizing that the metabolic foul-up was, in turn, due to the lack of a single enzyme. The gene, then, was responsible for ordering the construction of one of those massive, complex molecules known as enzymes—the compounds that act within the body to facilitate every kind of chemical activity from digesting dinner to synthesizing new DNA. The manner in which, quite literally, genes make the man seemed suddenly much clearer: One gene—one enzyme.

In years to come, the study of sickle-cell anemia would reveal that the defective red blood cells characteristic of that hereditary disease are due to a gene-linked protein deficiency, and the one gene—one enzyme theory would be expanded to one gene—one polypeptide, polypeptide being a general name for the complex molecules characteristic of both enzymes and structural proteins. The principle itself, however, continued essentially unchanged from Archibald Garrod's unheralded original observation.

By the 1940s then, genetic research was coming along rather well, considering the monumentally basic questions it had to deal with. But what about the first line of scientific inquiry we began with, the other half of our promised collision: Friedrich Miescher's nucleic acid?

By 1920, it was known that there were two types of nucleic acids, and the prevailing wisdom had it that one kind was found in animal cells, the other found in plants. In the decade following, the misconception was corrected and deoxyribonucleic acid (DNA) and robonucleic acid (RNA) were properly distinguished. But no one was quite sure what these curious molecules were good for—and many chemists, sticking cautiously to their own bailiwicks, quite frankly didn't care.

Some researchers suspected that the nucleic acids were involved with heredity—and by 1924, in fact, these substances had even been identified as the major component of the chromosomes. Yet oddly enough, the debate over the real function of those complex acids would continue right up to 1951, just two years before James Watson and Francis Crick introduced DNA into the common vernacular.

Most curious of all, however, was that the final collision between chemistry and genetics had taken place seven years before that, in the midst of World War II, with the discovery of the first real connection between DNA and the process of heredity, and this discovery was occasioned by the death of a rat. The rat in question died of pneumonia, in an elaborate experiment devised by three researchers named Avery, MacLeod, and McCarty at the Rockefeller Institute in New York, and based on a well-known but thoroughly mysterious property called bacterial transformation.

Decades earlier, a researcher had shown that if one took two different strains of the pneumonia bacterium —one of them a mutant incapable of causing disease, and the other a deadlier version that was, however, itself dead—and injected a mixture of the two into a mouse, the mouse would die. This was very odd: What, exactly, was killing the mouse? The live strain couldn't do it by itself, and the dead strain, almost certainly, hadn't come back to life. The only conclusion possible seemed to be that the killer characteristics of the deceased bacteria had somehow been transferred to the non-lethal mutants, which then not only became deadly themselves, but passed on that quality to their offspring.

Odder still was the discovery that this process of "transformation" would occur even if the mutant bacteria were merely exposed to a chemical extract derived from the deadlier cells. The question was obvious: What, precisely, was the nature of this transforming material?

Avery and his colleagues decided to investigate just how the mouse-killing capacity passed from dead bacteria to live. They gradually removed one chemical compound after another from the bacterial extract—first the proteins, then the starches and fats, until there was really nothing left floating in the mix except a minute, fibrous thread that Avery could actually pick out of the solution on a glass stirring rod. Redissolved, the material of that tiny thread transformed harmless bacteria just as well as the whole extract had in the first place. The material was deoxyribonucleic acid: DNA.

Our collision, clearly, happened in that New York laboratory: DNA was finally identified as the agent that is capable of carrying information. Yet, curiously, the notion didn't really take hold at first. It was almost as if, as the collision became imminent, some intellectual inverse-square law of repulsion came into play. In the years between the dead rat and the double helix, much experimental effort would go toward crediting *proteins* as the vehicles of heredity.

Why? A major problem was how mere molecules, in the first place, can carry information as complex as the shape of a fruit fly's wings, or the length of a sweet pea pod. By the 1940s, the nucleic acids had been recognized as mammoth molecules—assemblages of vast numbers of smaller molecules, hooked together by the invariables of physics and chemistry. Large as they were, however—larger, even, than their closest competitors, the proteins—they were nonetheless thought to be only a chattering repetition of the same four chemical constituents. And how could such mindless repetition ever read out physically in terms as complex as a tree—or a human being?

Proteins—a ubiquitous class of compounds positively rife in the single cell—looked in all like a much better bet. The long DNA molecule is, after all, composed of only four separate substances; proteins, on the other hand, are produced from a palette of twenty different amino acids. In retrospect, the whole notion of molecular genetics was really rather a flagrant grab at enlightened vitalism; and should the mysteries of life yield to mechanistic explanation, it seems only right that it should at least have been a *complicated* explanation.

The first indication that DNA might in fact be somewhat more complex than initially thought came in 1948, fully four years after Avery's historic work with bacterial transformation. A biochemist named Erwin Chargaff looked at DNA with a new technique called paper chromatography, a method whereby one forces the molecules of a given substance to seep along a dampened paper strip. By observing the relative speed with which the individual components of the substance move, and often the color of the layer formed by each of them, one can determine their identity.

Chargaff discovered very quickly that DNA was not a simple, repetitive chain—that in fact, the four principle components of the big molecule clearly occurred in unequal quantities depending on the exact source of the purified DNA. DNA from human sperm, Chargaff found, contained different proportions of the four components than did yeast cells, and DNA from the tuberculosis bacterium was constructed differently than both.

That discovery represented major credentials for DNA as the unit of heredity, but even the combination of dead rats and paper chromatography was not sufficient to carry the day for DNA. The research that finally established DNA's place in the genome came in 1952, and involved two new experimental creatures in the laboratory menagerie. These two creatures remain, to this day, at the forefront of molecular genetics, and thus will take starring roles in the new matter of artificial DNA recombination.

3

The Experimental Menagerie

The bacterium *Escherichia coli,* named after its discoverer, Theodor Escherich, and the portion of the human intestine that may well be its favorite habitat, is so small that several hundred thousand could fit comfortably on the period at the end of this sentence. *E. coli,* according to James Watson "is now the most intensely studied organism except for man."

Why is so much attention being paid to this tiny, rod-shaped microbe? For the same reason, ultimately, that it has been paid to sweet peas, fruit flies or bread mold: *E. coli* represents, for the researcher, an easily raised and rapidly reproducing population of genes available for the watching.

That hasn't always been the case. It wasn't, in fact, until the 1940s that science was even certain that bacteria *had* genes. All living cells, it seems, be they bacterial, human, or rubber plant, fall into two great categories: the prokaryotes and the eukaryotes. The prokaryotes seem to represent evolution's prototypical cell structure, a design that is still found in all bacteria and in blue-green algae. Prokaryotes have no nucleus binding up the genetic material, the nucleus apparently being first introduced on the later eukaryotic model, sometime during the Precambrian era.

All studies of classical genetics had been done with eukaryotes, like fruit flies or bread mold. Because these bacteria had no nucleus housing their chromosomes, no one was quite certain just how they handled their genetic affairs. As late as 1938, in fact, a major text on the history of bacteriology did not even include the word *genetics* in its index.

Well, it turns out that bacteria have genes too, arranged in chromosomes that float around loose in the center of the cell but which otherwise operate in a manner similar to the eukaryotes—similar enough, at any rate, to suddenly make bacteria fair game as laboratory animals for the study of genetics. By now, many molecular geneticists have several million laboratory helpers: a colleague or two, a couple of graduate students, a handful of technicians, and an immense colony of *E. coli*. Room is no problem: While *E. coli* is shaped rather like a Goodyear blimp, it measures at most only a few ten-thousandths of an inch in length. It usually thrives in fairly simple solutions of various salts and sugars, or broths extracted from meat or yeast. And *E. coli* is sufficiently hardy that viable samples have even been retrieved from frozen feces discovered in the long-abandoned hut of an early Antarctic explorer.

Perhaps the biggest advantage of *E. coli* lies in its over-whelming reproductive capacity: In the time it would take one of Mendel's pea plants to sprout, a researcher can observe several hundred successive generations of the tiny bacteria. In just twenty-four hours, one bacterium produces a population in the vicinity of one million billion—more than enough to guarantee a scientist the rich supply of mutants essential for genetic research.

That bacteria suddenly became an acceptable subject for such research was another reflection of the growing scientific certitude that the business of molecular genetics is a universal one—that the study of tiny bacteria living on sugar water in glass could yield insights applicable as well to human beings. That same genetic unity, however, as we will see, would become a central issue in the matter of deciding precisely how safe recombinant DNA was.

But *E. coli*'s contribution to molecular genetics might

never have occurred were it not for one more creature in the menagerie of molecular genetics: A virus so specialized that—in accord with the adage that even fleas have fleas—it specializes in living off bacteria.

In 1915 an English researcher named F. W. Twort noticed that something had gone curiously awry with a culture of bacteria in his laboratory. Their appearance in his shallow glass petri dish had changed from a normal creamy white to translucent. The bacteria were, moreover, dead as doornails: "lysed"—burst wide open —in the language of microbiology.

The affliction appeared contagious: A bit of the doomed colony that was transferred to a fresh colony produced the same result after a day or so. And this occurred even when Twort did his best to filter out every possible particle from the original culture—right down to the smallest bit of dead bacterium. Twort finally concluded that some separate infectious agent was killing his germs, and he tentatively fingered a virus. The whole notion of "virus" was then still very new, having only recently been noted in similar filtration experiments among plants. The virus itself, however, would remain invisible for decades to come.

Two years later and just across the English Channel, a researcher named d'Hérelle noticed exactly the same thing as Twort. This time, however, the experimenter was sufficiently certain of the odd phenomenon that he went ahead and named those invisible agents of bacterial destruction, calling them "bacteriophage"—literally, eaters of bacteria.

D'Hérelle, moreover, hoped that bacteriophages and their unusual appetite might be marshaled to combat bacterial infection in the human body. Two decades of research revealed, alas, that while bacteriophages quite happily assault and destroy immense numbers of bacteria in the laboratory, they are not nearly so reliable in the body. One explanation put forward suggests that the body's own immune system cautiously knocks out the bacteriophage before they even have a chance to act. At any rate, after two decades of work and the compilation of a great deal of data, the discovery of antibiotics prompted the abandonment of phage as a cure for infectious disease—but not before Sinclair Lewis immortalized the humble phage in his novel *Arrowsmith*.

But phage would again have its day. And that day would begin in approximately 1938, when Max Delbrück, a student of Neils Bohr and thus one of the first of that wave of physicists attracted to biology, decided that phage might be a particularly useful way to get at the mechanisms of genetics.

The notion itself was not new. Hermann J. Muller —the student of T. H. Morgan who stopped searching among endless numbers of fruit flies for natural mutations and started creating his own with X-rays—had earlier suggested that phage reproduction might be an interesting subject for study. The fact was that phage seemed to do almost nothing *but* reproduce, and were, moveover, the very smallest biological entities capable of that feat. A phage's genes alone can constitute fully half the total weight of the whole creature.

Both Muller and Delbrück were, as we will see, exactly right. By now, many phages have been identified, with various structures and more or less bizarre ways of getting along in the microbial world. But of special interest here are those which happen to infect *E. coli*. *E. coli* and its associated phages have been, in fact, for nearly a quarter century now, something of the Mutt and Jeff of molecular genetics.

So what is a phage? Essentially, it's a mass of DNA, encased like a chocolate-covered cherry in a geometric protein shell, with a tail that, when deployed, can look vaguely like the landing struts on a lunar module. When a phage infects an *E. coli* bacterium, as seen in electron micrographs, it almost literally squats on the cell wall, placing its geometric body right up against the membrane, and then injects a long squirt of its own genetic material, a lengthy, trailing DNA molecule that leaves behind an empty head—what the researchers themselves call a "ghost."

The injected DNA, now free to roam the interior fluid

of the cell, ultimately proceeds to reprogram the regular cellular business of the infected bacterium and convince it, instead, to create more phage. Those new phage, manufactured from the raw physical material within the bacterium, usually multiply at an enormous rate until they finally burst—lyse—the walls of the cell that initially gave them life. Those phages, once released, go on to colonize other nearby bacteria.

In a sense, the phage represents almost pure information; without hijacking some hapless *E. coli*'s cellular mechanism, the phage would remain only a bit of isolated, encapsulated DNA. Phage exist so near the molecular boundary of life that when the first electron microscope pictures of the tiny creatures appeared, researchers assumed that their tails were responsible for their motion. Later research revealed that not only do the tails strike the bacterium first, but that simple Brownian motion—an effect first observed in inanimate microscopic particles in the year 1827—accounts for the collision rate between phage and their victims.

Phage, then, are almost pure genes. And so it is little wonder that phage grew suddenly attractive to the new generation of molecular geneticists—so attractive, in fact, that within a few years after World War II the small circle of researchers who gathered around Max Delbrück called themselves the Phage Group. The Phage Group further spread the techniques and potentials of phage research during summer seminars conducted at a small scatter of laboratory buildings in a wooded area of Long Island called Cold Spring Harbor.

Cold Spring Harbor is by now both an exclusive preserve and a regular pilgrimage for molecular geneticists. It was at Cold Spring Harbor that, in 1952, with the help of many million bacteriophage the final evidence arose to clinch the association of DNA with genetic information.

Of course Avery, with his bacterial transformaion experiment in 1944, had made the same point rather firmly. But for a variety of reasons—from the dubious structure of the bacteria involved—resistance remained to the idea of such a role for DNA. Avery himself remained resolutely cautious: "It is lots of fun to blow bubbles," he wrote to his brother, "but it is wiser to prick them yourself before someone else tries to." The molecular biologist Gunther Stent in his book *The Coming of the Golden Age* describes the scientific unwillingness to accept the evidence for DNA with a wry quote from Sir Arthur Eddington: "It is also a good rule not to put overmuch confidence into the observational results that are put forward until they have been confirmed by theory."

Theory was stuck deciding between protein and DNA as the actual medium of genetic transfer. Phage—being composed of nothing *but* protein and DNA—looked to be a particularly apt test case. Clearly, whatever phage was squirting into *E. coli* must also be the the material that carries the information.

Thus, in 1952, amidst the luxuriant maples and dogwood of Cold Spring Harbor, two researchers named Alfred Hershey and Martha Chase were "literally forced" to do something now known as the "blender experiment." That experiment was both simple in concept and profound in its implications. It used some fairly new techniques to mark a crop of phages with radioactive isotopes: The protein coat was marked with radioactive phosphorus. The marked phages were then mixed in solution with bacteria, which they promptly infected. The infected bacteria were then spun for several minutes in a blender, which threw the phage particles off the surface of the bacteria they'd attacked. But the "transforming material"—its identity then still in question—was already in the bacteria. The whole mixture was then transferred to a high-speed centrifuge, which separated the solution into two portions: a tiny pellet of compressed, infected bacteria, at the bottom of the test tube, and the remaining fluid, in which floated the far lighter, now empty, phage ghosts. Two distinct portions of the centrifuge run were analyzed for their individual radioactive markers and the results are history: The phage protein was left floating in the form of the ghosts; the phage DNA was already safely and securely inside the bacteria.

The "transforming principle" had once again been tracked into its lair, and now there could be little doubt as to the genetic credentials of DNA. And then just one year after the blender experiment came the event that still officially serves to mark the birth of molecular genetics: Watson and Crick's description of the shape of the DNA molecule itself.

That Nobel prize-winning description was the culmination of a breathless scientific competition that James Watson himself aptly characterized in his book *The Double Helix*. And it is a competition that continues to this day in retrospective evaluations of each participant's real research contribution.

Little wonder: The double helix description of the DNA molecule—a winding spiral staircase of two long chains of sugars and phosphates, linked by rungs of four different organic bases—is by now almost as well recognized a public image as is the electron-encircled atomic nucleus and is, moreover, as significant in symbolizing new comprehension of the underlying mechanisms of nature.

"A structure this pretty just has to exist," is the way one original researcher put it. And in fact, the helical nature of the molecule served immediately to answer some of the longheld objections to casting DNA in the role of genetic medium. Watson and Crick concluded their brief paper in *Nature*, in April 1953, with what Gunther Stent called "one of the most coy statements in the literature of science." "It has not escaped our notice," Watson and Crick wrote then, "that the specific pairing we have postulated suggests a possible copying mechanism for genetic material."

Things had clearly moved along rather briskly in the half-century since biophors, gemmules, and stirps. And the final, firm identification of the molecular nature of heredity, along with the deciphering of its novel doubly helical structure, launched two more decades of yet more intense molecular biology—a sequence of discovery that led directly to the moratorium on recombinant DNA in the spring of 1973 and culminated, eight months later, in the conference at Asilomar.

The decades between had concentrated on the larger question of molecular genetics: How does any molecule, complex and varied as it may be, direct the construction of something as complicated as a human being? Or as complicated, for that matter. as *E. coli?*

That question won't be answered in this book; nor, one suspects, in any other book for some time to come. Nevertheless, much has, in fact, grown clearer over the decades since the double helix.

The DNA—the genes—within a cell directs the assembly of those two basic organismic constituents common to every creature from bacteriophage upward: enzymes and structural proteins. The specific chemical information is contained within the DNA molecule in terms of the sequence of *nucleotides*—the individual rungs—on that sugar-phosphate ladder. The unraveling of that code was a major achievement of the late 1950s; in a broad sense, it was the final explanation of how an inanimate molecule might contain the descriptions for a whole array of life-substances.

DNA, however, does not do the job by itself. That second group of nucleic acids, first misinterpreted and then finally identified in 1926 as ribonucleic acid or RNA, remained in experimental limbo until the late 1950s. In the years since then, it has grown clear that not only is RNA a biochemical helpmate to DNA but that there are, in fact, three different kinds of RNA within each cell.

Messenger RNA is the variety that comes from the segment of the DNA molecule that codes for a given enzyme or protein; it takes the data obtained from the DNA through the fluid interior of the cell to a tiny, sub-cellular unit called a ribosome—the cell's factory, so to speak. The other two varieties are ribosomal RNA, which resides within the minute ribosome, and transfer RNA, which is found in the inner fluid of the cell. Transfer RNA carries the necessary raw materials from the outskirts of the cell to the ribosome, which then manufactures the particular enzyme or structural protein asked for by the original DNA segment. This process is repeated over and over again by the three kinds of RNA working together—as blueprint, supplier, and factory— according to varying instructions by the DNA until a single cell has been produced.

The preceding paragraphs summarize decades of intricate biochemistry and painstaking deduction, and don't even begin to suggest the real complexity of the process itself or the magnitude of the questions that still remain.

How, for example, does a set of cells know how and when to *differentiate*, to stop simply repeating themselves over and over and begin, with marvelous subtlety, to become the blend of tissues that is, say, a bicep? And how does some facet of that same control process run amok and yield not a bicep, but a tumor?

These questions involve the most fundamental processes of molecular genetics. And unfortunately, when DNA, in the course of its most intimate functions, has been laid bare, it has also, rather firmly, stopped functioning altogether.

Until recently. But that, for our purposes, is another story altogether.

4

The Moratorium

Our miniature portrait of molecular genetics, having reached that collision between chemistry and genetics, is now about to resurface at a crucial juncture in that process.

Much of the work in molecular genetics since the description of the double helix concerned the internal molecular processes that result in specific biochemical products. The remainder dealt, essentially, with mapping the genes themselves—complex elaborations of T. H. Morgan's initial work with *Drosophila* that by now have offered up, for a handful of bacteria, elaborate descriptions of just what each segment of a given genome's DNA might encode.

Yet all that still suggests a certain separation between chemistry and genetics, and also a limitation inherent in the theoretical nature of the research and its results. Until recently the work done has been mostly painstaking dissection with little impact other than intellectual on the world outside the laboratory—"until recently," because now a new biochemical tool has been added to the workshop of the molecular geneticist.

"Restriction enzymes," James Watson told me six months after Asilomar, "are one of the most important tools to come along yet." Restriction enzymes are a class of chemicals distilled from dead bacteria—enzymes that those bacteria used in life to protect themselves against

41

the invasion of foreign DNA, in the form of, say, bacteriophage. Restriction enzymes cleave the foreign DNA molecule into segments and thereby render it noninfectious. This biochemical meat cleaver in nature has rapidly turned into a scalpel in the hands of molecular geneticists.

It's never good sense to single out one particular discovery or technique as the turning point in a discipline; even the now mythic deciphering of the double helix was, in the broader view, an inevitable incident in the complex history of molecular genetics. But, for what the use of restriction enzymes has already achieved—and the technique, now known as recombinant DNA engineering, is still in its infancy—it has earned a special place in that history.

Restriction enzymes cut DNA molecules with such precision that they would likely have made Thomas Morgan give up his microscope forever. No more peering at blurrily defined chromosomes; restriction enzymes allow one to dismantle DNA so neatly that one can actually begin to map the molecular components of the individual genes themselves. More importantly, it also allows one to *reassemble* pieces of altogether different DNA—genes from creatures as disparate as the researcher can manipulate—into new, hybrid molecules that by all indications should function just as well as do natural genes.

Beyond its purely scientific significance, the technique of recombinant DNA engineering has played a unique role in the brief history of molecular genetics: It has triggered front-page headlines and public furor over "genetic engineering," has shut down research, has raised worldwide controversy, and has brought about, finally, the first international assembly of molecular geneticists convened solely for the purpose of deciding what, precisely, to do next.

It all started, quietly enough, at the Stanford University Medical Center, on the thousands of acres of prime mid-California farmland just south of San Francisco, settled by a strong-willed California governor named

Leland Stanford and turned into a university by his equally strong-willed wife. Precisely as did its founding family, Stanford University tends to get what it wants. And when in the late 1950s Stanford decided that it needed a biochemistry department to go along with its medical school, it managed to assemble a very good one, headed by Nobel laureate Arthur Kornberg.

Kornberg had been responsible for some of the most exciting biochemistry in the decade following the description of the double helix, particularly the "cell-free" laboratory synthesis of nucleic acids. When Kornberg moved to Stanford from Washington University in St. Louis, he brought with him a whole set of researchers and even some hardware. Stanford had, in effect, purchased a biochemistry department intact, and had chosen with sufficient wisdom that it is still considered one of the top biochemistry departments in the business.

A Brooklyn-born researcher named Paul Berg was one of the young scientists involved in that westward migration. Berg had already won recognition for his work with Kornberg on how *E. coli* synthesizes its proteins; by 1965, he had set up chemical models that could produce both proteins and RNA in glass—itself a complex feat of biochemistry.

Berg had been at Stanford about eight years when he became curious as to whether the same processes of gene expression and protein production operate in similar manner in the higher organisms—in, for example, cells from mammals.

Unfortunately, however, once one moves to the arena of the more complex eukaryotic cells, the trusty bacteriophage is no longer available as an experimental probe. Berg suspected that a reasonable counterpart would be the tumor viruses. Just as phage do in bacteria, tumor viruses can insinuate their spurious genetic blueprint into the chromosome of the mammalian cell, and thus take over the cellular machinery. But occasionally, instead of turning out countless identical copies of themselves, tumor viruses order the cells they have infected to reproduce themselves, in that uncontrolled pattern of growth we know as cancer.

These viruses, then, looked as if they might offer as many clues to the genetic secret of higher cells as phage had offered for bacteria. And so Berg spent a Stanford sabbatical studying animal viruses in Renato Dulbecco's laboratory at the Salk Institute in La Jolla, California, an early center for tumor virus research. After a year at the Salk Institute, Berg returned to Stanford, convinced that "this was the field we should go into in depth."

Within a few years, bacterial systems had nearly been phased out as research subjects in Berg's lab. A new set of postdoctoral students had been drawn to the study of animal viruses; lab space expanded and the possibilities inherent in the genetics of tumor viruses looked very promising.

That is, until 1972. It was then that Berg and several students began work on an experiment using a tumor virus that, for the first time in his career, he eventually decided was too dangerous and so called a halt to the experiment. It was the first time for Berg and, quite likely, the first time for molecular genetics as a whole—but not, by any means, the last.

In order to understand the experiment that Berg canceled, it is necessary to introduce yet another character in the menagerie of molecular genetics—a type of virus known as Simian Virus 40. SV40 is one of a handful of animal viruses that have, in recent years, become as central to the study of viral genetics as *E. coli* has been to bacterial genetics. SV40, however, has a history not quite so benign as the ubiquitous *Escherichia coli*.

SV40 was first isolated from the kidneys of rhesus monkeys in 1960—a batch of kidneys that had been destined for use in the production of the new poliomyelitis vaccine. The virus had escaped notice for years previously as it caused no apparent disease in the monkeys themselves. Injected into newborn hamsters, however, it caused tumor growth; and it did the same, moreover, to human cells in test tubes.

SV40 attracted instant attention because by the time

it was discovered, vast amounts of polio vaccine had already been produced in monkey kidney cultures rife with SV40, which in some cases survived to inhabit the vaccine itself. Thus, from ten to thirty million American children in the years between 1955 and 1961 received, along with their polio shot, a dose of live SV40 virus.

By now, considerable discreet medical surveillance has been directed at the known SV40 recipients—and thus far there has been no evidence of any mass onset of malignancy which, considering the numbers involved, could have made the thalidomide tragedy look insignificant in the annals of self-inflicted human suffering. Some recent studies, however, have suggested the presence of SV40 in association with some human neurological diseases that involve progressive central nervous system degeneration—a class of disease not included in the original epidemiologic surveys of SV40 infectees.

Those unsettling findings, combined with increasing medical appreciation of just how slowly some viruses act, have kept the books open on the absolute safety of SV40. If science has learned anything about viruses thus far, it is that they are exceedingly tricky. That knowledge, however, hasn't kept SV40 out of the laboratory—or, for that matter, out of researchers' bloodstreams. SV40 is by now a laboratory favorite, and many who work with it rapidly develop bloodstream indicators that the virus has infected them. But thus far those blood tests are the only tangible proof that SV40 has paid a visit. It seems a small price to pay, considering what certain other monkey viruses can do to a human being —and a small price also in that a fairly tractable, tumor-producing virus is just too interesting an experimental creature to let go.

SV40 had been a standard in Paul Berg's Stanford laboratory since his year at the Salk Institute. And thus it figured centrally in that first abandoned experiment: an attempt to learn whether SV40, with its ability to subvert the metabolic apparatus of animal cells, could be used as an experimental probe in the same way bacteriophage is used with bacterial cells.

Berg designed an experiment to test the capabilities of SV40 as a vehicle to carry animal genes into animal cells. But since, at that point, animal genes were not available to use as passengers, Berg chose instead some bacterial genes from a tiny phage called lambda dv.

In the early 1970s, joining two different DNA's together outside the living cell represented a bioengineering task approaching impossibility. But the potential rewards, in terms of knowledge, were so great that two laboratories at Stanford set out to devise a technique—Berg, Bob Symons, and Dave Jackson in one; Peter Lobban and Dale Kaiser in another. Before long they had developed an elaborate biochemical process, utilizing rare and costly enzymes, that promised to make the notion a reality—to cut the phage DNA apart, to cut the SV40 DNA also, and then to reassemble them into a single, composite molecule, combining genes from both the virus and the phage.

It was soon recognized that the resultant hybrid molecule might very well be able to go both ways—that the SV40 segment was capable of carrying the phage DNA into animal cells, and that the phage DNA might work just as well at carrying the tumor virus DNA into bacteria. Although the latter possibility had not even been the intent of the original experiment, it was a sufficiently scary possibility that the work planned with the newly constructed molecular hybrids shut down in short order, and it remains canceled as of this writing.

Why? "While I didn't think of it exactly this way at the time," Berg says now, "I must have realized that I'd been wrong many, many, times before in predicting the outcome of an experiment, and that if I was wrong about my assessment of the risk in *this* experiment, then the consequences were not something that I would want to live with."

That ominous assessment began in 1972 when one of Berg's associates, a new graduate student named Janet Mertz, described the proposed experiment at a tumor virus workshop held that summer at Cold Spring Harbor. Mertz had the task of developing techniques to facilitate the uptake of DNA by the subject *E. coli* cells. But when, that summer, she described an experiment in which *E. coli* would be urged to absorb SV40 DNA, there was immediate and vociferous criticism.

Long-distance telephone calls ensued immediately between the Long Island research retreat and Berg's lab in California. "'My God,' people said, 'you can't put SV40 into *E. coli*!'" Berg recalls. "I think I was upset by the criticism at first, but then I went out and started to talk about the problem with a lot of people."

The question, quite simply, concerned the wisdom of transplanting SV40 genes, conceivably coding for tumor production, into a bacterium that not only lacks that capacity to begin with, but which lives in virtually every human gut on the planet.

The answer was not simple. No one could be certain what that combination of genes might do, and the possibilities ranged from nothing at all to some nightmarish version of contagious cancer. No one knew because the combination simply didn't exist in nature. Two years later at Asilomar, a British biologist would put it very neatly: "Evolution is our only experiment." But the results of that long-term experiment are peculiarly ambiguous: While there can be little doubt that nature has for eons done its best to thoroughly mix its gene pools, it's not nearly so clear as to whether the contemporary absence of a given genetic combination means that the blend has been tried and discarded in ages past —or whether, in fact, it represents something altogether novel and unreckonably hazardous.

There is still no real answer to that question, and thus Paul Berg, the first to confront the situation, was initially rather dubious. But Berg is not the sort to ignore misgivings, and in the months following the summer of 1972 he started asking his own questions.

In the course of that questioning, Berg was already consulting the principal group of individuals who would organize both the moratorium and Asilomar. David Baltimore, a specialist in RNA viruses who would become, a few months after Asilomar, an unusually young

Nobel prize winner, was initially critical of the SV40 experiment. A researcher at the National Institutes of Health named Maxine Singer—an outgoing scientist about Berg's age and, like him, Brooklyn-born—was also uncertain. She and her husband Daniel, a lawyer with a special interest in scientific ethics, spent an evening at their Maryland home discussing the issue with Berg.

"We talked at great length about the ethical basis of science," Berg recalls, "and how this experiment fitted into that context, and again I was unconvinced. But then the more I talked to people, and the more I thought about it, I realized that while I could argue that the probability [for hazard] was low, I could never say that the probability was zero." Even if the tumor virus genes did not actually operate, or "express," within the *E. coli*—which would make the experiment, essentially, a failure—Berg realized that the modified bacterium could still be a carrier of latent tumor virus genes that might eventually enter an animal cell where they *could* be expressed.

"So we might be doing an experiment," Berg reasoned, "that would have no payoff, but which would be, at the same time, just as risky."

Berg canceled the experiment. "And then," he recalls, "the people who had initially opposed it, said, good, great, nobody else is going to do it if you're not going to do it." At that point, in fact, the molecular joining technique that the Stanford group had designed was so complex and sophisticated that chances were no other lab could have managed it even if they'd wanted.

And there—with Berg's voluntary deferral—the question of introducing potentially hazardous genes into *E. coli* seemed to die aborning. The only real result of the ethical turmoil was a conference on the relatively new question of biohazards, held in January 1973 at the Asilomar Conference Center. The biohazard question then centered solely around the tumor viruses themselves, with no hint of the far more devilish complexities that the same Asilomar chapel would hear just two years later. Yet even so, as Berg recalls, "At the end of the biohazards meeting we were supposed to plan a second meeting, but we were so exhausted, and there seemed so little hard information on whether tumor viruses were dangerous, that the idea of a second meeting was finally dropped altogether."

But only for the moment. Because at precisely the same time, research was under way at Stanford that would abruptly have far more wide-reaching consequences than did Berg's first, deferred experiment. The experiment with SV40 had used one particular variety of the newly discovered restriction enzymes to cut the viral DNA molecule into fragments. That restriction enzyme, first isolated by a soft-spoken San Francisco biochemist named Herbert Boyer, was called EcoR1—Eco, because it was derived from *E. coli*.

But EcoR1, it soon developed, had a very special capability that no one who used it had even dreamed of. It seemed that SV40 DNA molecules that had been cut by EcoR1 and inserted into monkey cells seemed to rejoin themselves spontaneously into a circular and functional bit of DNA. This was not the way it was supposed to be: As nearly as anyone could tell, restriction enzymes, being a defensive reaction of bacteria, should take foreign DNA molecules apart in such a fashion that they could never come back together again—at least not without the elaborate biochemical help that Berg's group had designed into his canceled SV40 work.

Yet the SV40 DNA molecule, when cut by EcoR1 and then placed into an animal cell, seemed to act pretty much as though nothing at all had happened. "That is," says Berg, "the DNA was completely infectious, even after cutting—and that was totally unexpected."

Berg suggested to Janet Mertz that she look into this paradox. While Berg was away on a month-long trip to Spain, Mertz and Ronald Davis, a new assistant professor at Stanford, discovered something altogether unexpected: The SV40 DNA, cleaved by EcoR1, was able to "recircularize" (in effect, rejoin itself chemically) without any of the elaborate biochemical assistance that had been used earlier.

"They're never given credit for it," Berg says of Mertz and Davis, "but in fact they did the first experiment which showed that covalently joined hybrid DNA molecules could be made in the test tube."

Berg and his colleagues called Herbert Boyer in San Francisco—who had, nearly a year earlier, sent the EcoR1 down to Stanford—and told him what the enzyme seemed to be doing. Boyer rushed down the peninsula to Berg's lab and verified the fact that the enzyme cut DNA in such a way as to allow the ends to be rejoined spontaneously, providing the potential ability to combine genes artificially from different species that could never mate in nature. In the months following that discovery, the Asilomar Conference on Recombinant DNA Molecules rapidly became inevitable.

But how does one make genes—separate segments of DNA molecules—unite artificially? EcoR1, it developed, was in fact a very special restriction enzyme. As do all its fellows, EcoR1 cuts DNA molecules into pieces. But EcoR1 does something more: it cleaves the double helix in such a fashion that "sticky ends" remain, resulting in a brand of biochemical Lincoln Logs that, when brought into close proximity, rejoin under the compulsion of fundamental electrochemical forces—and rejoin so completely that they then operate just as well as natural genes.

There was no immediate explanation for EcoR1's curious properties. Even its discovery had been something of an accident. But the implications of EcoR1 were very clear: The exceedingly difficult experiment that Paul Berg had deferred was now far less difficult—and it could be done, in fact, by anyone with moderately refined laboratory skills.

The second development which insured the rapid acceleration of recombinant DNA work was even then under way, in Stanford's Department of Medicine. At just about the same time that the curious new properties of

EcoR1 were being discovered, a young researcher named Stanley Cohen began to suspect that his specialty —a peculiar cellular hanger-on known as a plasmid— might in fact be an apt vehicle for propagating foreign genes.

A plasmid is a tiny bit of extrachromosomal DNA that floats rather freely within the saclike cell of many simple bacteria. Like bacteriophage, plasmids represent almost pure genetic information. But while phage, safely encased in their protein shells, can afford to traverse the vast lonely spaces between bacterial hosts, the naked plasmids are far more vulnerable. And thus to insure their survival, plasmids tend to be benign and even helpful bacterial guests, performing fairly specific genetic favors for their hosts in return for continued residence privileges.

Stanford researcher Joshua Lederberg earned his Nobel prize for describing the helpmate role that plasmids play in the curious, asexual process of bacterial "mating" called conjugation. A decade after that, plasmids were implicated as well in the rapid, unsettling rise of bacterial strains that have gained resistance to antibiotics.

Plasmids, thus, already had a number of established roles by the time Stanley Cohen tried to use them as potential vehicles for carrying bits of foreign DNA into *E. coli*. It was a natural combination—the plasmid involved had originally been derived from the same sort of bacterium—and the experiment was so successful that almost immediately the minute plasmid had gained another role. Within months, recombinant DNA techniques were already referred to in newspapers as "plasmid engineering."

The combination of EcoR1 and the facility with which plasmids could be grafted onto the recombinant DNA process meant that the dilemma that had been Paul Berg's exclusively in 1972 now suddenly belonged to science at large. Not long after the first work with plasmids, a group consisting of John Morrow, Herbert Boyer, and Stanley Cohen demonstrated that DNA from the African frog *Xenopus* could be stitched into a plasmid and subsequently replicated within *E. coli.*

The Morrow group's experiment underscored the potential power of the new techniques, and the field was suddenly wide open. With the exotic enzyme, willing plasmids, and a moderate level of microbiologic skill, one could perform potentially hazardous genetic manipulations that two years earlier had not even been imaginable. It took no imagination, however, to formulate what suddenly became the most important question: Precisely what was science going to do about it?

Herbert Boyer, the biochemist who had isolated EcoR1 in the first place, had come to report the first experiments with hybrid DNA molecules at the 1973 Gordon Conference on Nucleic Acids, held in New Hampshire. Maxine Singer, the NIH researcher with whom Berg had shared his misgivings two years earlier, happened that year to be cochairman of the conference. Following Boyer's description of how EcoR1 might be used to construct new plasmids, Singer recalls one conference attendee observing, "Well, now we can put together any DNAs we want to." "It was that remark," Singer says, "that everybody heard."

The next morning, Singer brought the question of the safety of recombinant DNA work up for consideration. In an unprecedented move, seventy-eight of the ninety-odd participants voted to send a letter of concern to the prestigious National Academy of Sciences. And then a far closer, and ultimately more critical, vote of 48–42 called for wider publication of that same letter.

"We are writing to you," as the letter finally read, in a September 1973 issue of *Science*, "on behalf of a number of scientists, to communicate a matter of deep concern." The matter wasn't even then yet known as recombinant DNA—it required fully thirty highly technical magazine lines just to describe the problem. But the implication was obvious: It was the duty of the NAS to study exactly what should be done about the public health implications of this abrupt and unprecedented alteration of the human relationship with the orderly process of evolution.

Even that early, Singer remembers the generation gap that would mark the entire course of the proceedings. "The younger people, some still graduate students, others post-doctoral fellows, brought to the discussion a whole context of concerns about public knowledge and social responsibility." Singer's own role as cochairman had been significant as well; she shared some of those concerns, and later she would reflect that had one of the more conservative members been chairing the conference, "the whole thing wouldn't have gone to first base."

It went, of course, far past first base. "Do you really want to publish this?" Philip Abelson, an editor at *Science*, asked Maxine Singer when she turned up with the Gordon Conference letter.

The answer was a foregone conclusion. The letter appeared forthwith, and within a matter of weeks, Paul Berg found himself once more embroiled in a controversy that he thought he had shelved nearly two years earlier. Quite logically—and partly at the suggestion of Maxine Singer—the National Academy called on Berg to ask him to help decide precisely what should be done about this newly troublesome matter of recombinant DNA.

Berg said he'd have to think about it. During a trip East in the early winter of 1973, he had lunch with James Watson, who had earlier, at the first Asilomar meeting on animal virus biohazards, been fairly adamant about safety issues—even saying that he would go to court and get an injunction to prevent workers in his labs from doing work that he thought dangerous. Berg and Watson agreed at lunch that it would be useful to call a meeting of people in the field and find out what sorts of experiments with recombinant DNA were being planned, and whether in fact those experiments were dangerous.

In the months since the work with EcoR1 and plasmid engineering, Stanford had been deluged with requests for the tiny pSC101 plasmid that was, at that point, the biological linchpin of the new technology. "The telephone calls were coming into Stanford daily," Berg recalls. " 'Send us pSC101.' 'What do you want to do?' we'd ask. And often we'd get a description of some kind of horror experiment and you'd realize that the person hadn't really thought about the possible consequences at all." Some of the nightmarish possibilities that actually occurred to researchers were work involving the potentially tumor-producing herpes virus, or experiments with diphtheria toxin genes—all to be conducted within that ubiquitous human companion *E. coli.*

Berg decided, with the blessing of NAS, to convene a small group of scientists to consider a response to the letter published in *Science.* That meeting, at MIT in in April 1974, was really an informal gathering of about ten individuals—Berg, Watson, Baltimore, along with several others who would soon become important figures in the controversy. Their first decision was easy: Another meeting, much larger, must be held to discuss the issues. Before leaving California for MIT, in fact, Berg had already inquired about space at the Asilomar conference center, a site both pleasant and fairly inexpensive, for the following February.

The second question was more difficult: What to do in the months between the MIT gathering and the Asilomar meeting? The pressure to go forward with the potentially revolutionary research was such that the group feared that by February, Asilomar would become "a 'show-and-tell' exercise"—a description of the construction of strains or hybrids of the very kind the group was most worried about.

"I remember very clearly," Berg says, "that it was Norton Zinder"—a Rockefeller University professor—"who said, 'If we had any guts at all, we'd tell people not to do these experiments.' That came as a shock. It seemed rather radical."

How radical it really was did not concern the group for long. The attendees at MIT were themselves among the most likely to want to do the work in the first place, and that seemed to make the notion of a moratorium more palatable. Ironically, it was precisely that same insularity of the decision-makers that would later draw criticism from those outside the field.

But at the time it seemed to facilitate the decision, and so a plan was set for another letter in *Science*—this time asking, in no uncertain terms, for a temporary moratorium on what looked like some of the most exciting biological research of the century.

It was not an easy request to make, and so over the next few months the letter went through numerous versions, starting from an original draft authored by Richard Roblin, an assistant professor of microbiology at Harvard Medical School. At some point, the word "International" was appended to the title of the proposed conference at the suggestion of Arthur Kornberg. "He said we shouldn't be misleading ourselves as to the impact of the letter," Berg recalls. "This was going to create quite an effect, and if it didn't apply to other people as well, there would be resentment." Four more signatures were added —California researchers who had been responsible for much of the initial work on the new techniques. And the letter was later amended to make it clear that it was an official National Academy of Sciences committee report.

During July 1974 the letter appeared almost simultaneously in issues of *Science*, *Nature*, and the *Proceedings of the National Academy*. Until that point, Berg was not struck by the moderately revolutionary nature of the enterprise. "Certainly it never occurred to me that anybody was doing anything historic," he says, "and I certainly didn't anticipate the kind of uproar that the newspapers gave it." The uproar started early. A curious *New York Times* reporter turned up with an early draft of the letter, apparently leaked by a student of one of the original signers. The reporter was put off successfully, but Berg admits to both surprise and reluctance when he was asked to attend a press conference to be held at the National Academy building in Washington on the day the moratorium letter was released to the public. "We hadn't thought that the public would be in on it," says Berg, "and I must admit I didn't give much thought at all to the press." Perhaps no one really did; although Berg now clearly recalls an observation by David Baltimore just before the press conference: "We may," he said, "have to live with this thing for a long time."

The Washington press conference, on a hot July Thursday in 1974, was attended by fifty or sixty reporters and was the first, in fact, that Berg had ever participated in—although it would not be the last. The Stanford researcher was surprised at how benign it turned out to be. The questions seemed neither penetrating nor even very interesting. "A lot of the reporters," Berg recalls, "were very uninformed. They just had no base or knowledge on which to ask questions."

The story, however, got both wide and dramatic play. Berg says, "When I saw the headlines—SCIENTISTS URGE BAN ON GENETIC RESEARCH—I realized that what we had done was being misconstrued." Much of that was due to genuine confusion on the part of reporters about exactly what sort of research was involved; the rest could be attributed to a slight but discernible trend toward unwarranted alarmism. Both reactions would color some of the media coverage for two years to come, even when the reporters involved had covered the issue since the outset.

After Berg's return to California, a whole parade of reporters began to materialize at his Stanford lab. The BBC interviewed him by telephone; the Australian Broadcasting Company set up a hookup with the campus radio station. A few months later, he even traveled to England to appear on a ninety-minute TV show called "Controversy." By then, the moratorium unquestionably qualified for that title. The version of the letter that had appeared in the English science journal *Nature* had dropped, due to the vagaries of transoceanic telex, a critical line from an early section, the lack of which made it sound as if the authors were suggesting the cessation of work with naturally occurring drug resistant plasmids—work which had been going on for at least a decade. The notion drew angry howls from English researchers like Ephraim Anderson—a world-recognized authority on the antibiotic resistance problem and Berg's debating opponent on the "Controversy" broadcast.

Years earlier, Anderson had been instrumental in the far-sighted British ban on the indiscriminate use of antibiotics in cattle feed—a practice now suspected to be responsible for creating a vast new reservoir of antibiotic-resistant bacteria. The misprinted letter, then, struck the British researcher as slightly presumptuous, and he fired off his own rather scathing reply to *Nature*. It was not until the BBC broadcast itself that Berg had the chance to ask Anderson whether he disagreed as violently with the moratorium statement as originally written.

The question was important, because the moratorium had been adopted in Great Britain with a zeal seen nowhere else. The Medical Research Council had declared all of the experiments described in the letter virtually illegal; even Berg would later express some misgivings about the wisdom of that response. But as we will see, Great Britain in 1974 was just recovering from what may well have been the first civilian biohazard accident in history: an escape of smallpox virus from a lab at the London School of Hygiene and Tropical Medicine that killed a young couple before it was tracked down and isolated.

The British, then, had a fresh memory of precisely how painful a natural biohazard can be; that they reacted so vigorously to the notion of artificially created biohazards is understandable. Only weeks after the press conference in Washington, a Working Party under the chairmanship of Cambridge Master Lord Ashby was charged with assessing both the potential benefits and the potential risks of recombinant DNA.

The controversy was only beginning. Six weeks after the NAS letter, *Nature* published an editorial statement asking the question "Should we publicise those experiments?" and expressing sympathy for the notion of moratorium. They would not, however, "commit ourselves to any blanket policy of turning away reports from those who work on regardless"—a clear reply to a suggestion by David Baltimore that editorial boards might find it appropriate to refrain from publishing papers from experiments covered by the moratorium.

The journal was likely wise in rejecting any putative policing role at that point. Researchers were trying, with less success, to do the same. The moratorium was a central topic of discussion that autumn; one international meeting, at Davos, Switzerland, brought forth observations ranging from "We are in the same place as nuclear physicists were in 1938" to "You should not hamper basic science; you cannot slow down research."

By January 1975, one month before the conference, pressure was intense. Yet the moratorium was almost universally observed in the eight months between publication of the letter and the first sunny Sunday at Asilomar. Maxine Singer recalls how pleased she was in the summer of 1974 at receiving a shipment of commercially produced restriction enzymes that carried a newly added cautionary statement that the material was sold with the understanding that it would not be used in the kinds of experiments proscribed by the letter.

Late in January the Ashby report was published in Great Britain, and press reaction was immediate. The headline in the following week's *New Scientist* put it simply: "NOT GOOD ENOUGH." The report laid strong emphasis on the potential benefits of the new technology and suggested, with only a few caveats, that the standard procedures used with known pathogens should be sufficient with recombinant DNA as well. The attention attracted by the London smallpox accident was cited as an example of just how rare the failure of standard containment techniques really is.

But then the Ashby group had not been requested to make recommendations as specific as the National Institutes of Health later demanded in the United States. And by the time the U.S. got around to real guideline writing, the British had launched a renewed attack on the matter that made even the elaborate NIH efforts look half-hearted.

During the months between the publication of the letter and the conference itself, now set for the last week in February 1975, Berg concentrated on details of organization. At a second MIT meeting, an organizing committee had been established consisting of Berg, Baltimore, Singer, Roblin, and two Europeans—Niels K. Jerne of Switzerland, then head of the European Molecular Biology Organization, and Sydney Brenner, a central figure in molecular genetics on the Medical Research Council staff in Cambridge, England. Brenner was a controversial choice at the time; a brilliant scientist, he could be tough and scathing as well, and some feared his possibly domineering presence. Months after Asilomar, however, Berg described the inclusion of Brenner as "the best decision we made."

The rest of the second MIT meeting was devoted to selecting chairmen for the individual committees, which by then had been narrowed to three interest groups: animal virus DNA; animal and plant cell DNA; and bacterial plasmids and phages.

Each committee would submit suggestions for participants. Maxine Singer's husband, Daniel—a lawyer and vice-president of the Institute of Society, Ethics, and Life Sciences in New York—was drafted to draw up a list of participants from the legal profession. Berg and the organizing committee selected the remaining few, particularly individuals from industry. Early on, it had been recognized that the industrial use of recombinant DNA—should it prove practical, say, for pharmaceutical production—would represent by far the largest-scale efforts in the field and therefore possibly the greatest risk of a hazardous new material escaping into general circulation. Berg invited five or six researchers, mostly associated with the pharmaceutical industries; the assessment of early industry involvement began to look accurate when another researcher—this one associated with General Electric—wrote to ask to attend the conference at Asilomar.

Already the Asilomar attendance list was growing rapidly. The organizing committee felt that it was incumbent upon them to create a representative group and yet one that was also small enough to operate effectively. Inevitably, dark mutterings about "intellectual lockouts" and the like were heard.

Some of the loudest of those mutterings, however, came from the direction of the press. Berg had wanted at most eight or ten reporters, with the proviso that they must spend all four days at the conference and not file until the conclusion; Daniel Singer was briefly charged with organizing that aspect of the operation. But press interest was intense, reporters were suspicious about dealing with someone not usually connected with press relations, and fairly quickly Singer turned the whole matter over to Howard Lewis, a low-key press officer for the National Academy who ultimately resolved the press crush by recommending that all sixteen applicants be allowed to attend. The requirement for four days' attendance proved finally to be both an acceptable and an eminently sensible sort of press filter.

The weekend before Asilomar was hectic. Each of the three committees were to have statements drafted for the full conference to consider. The eukaryotic DNA and plasmid people were meeting at Stanford to write their statements; the virus group was already at Asilomar. "Stanford was teeming with these guys," Berg recalls that last weekend before the conference. Sleepless nights and chronically defective photocopying machines hounded the two groups before their departure for the two-hour drive to the Monterey peninsula. "They looked like hell when they arrived," Berg says.

But the weather had just taken a turn for the better; the nearly 150 Asilomar participants arrived in the middle of a superb midwinter California coastal high pressure zone. And found themselves, right off, in the midst of another zone at least equally pressurized.

5

Meeting at Asilomar

The molecular biologists invited by Berg and company descended upon California's Monterey peninsula on very nearly the same day as did the monarch butterflies. The Asilomar Conference Center is a scatter of rustic dormitories and meeting halls hidden in a seaside forest of redwood and Monterey pine, just outside the tiny town of Pacific Grove. Traditionally, each February immense flocks of migrating monarch butterflies numbering in the millions briefly cover the trees there in thick sheets of orange and black.

In late winter of 1975, along with the monarchs, came the molecular biologists, shuttled in from the local airport, often still clad in overcoats donned hours earlier in Cambridge or Krakow. The weather that Sunday was crisp and the sky bright blue—the finest February climate that the Monterey peninsula can offer—and it seemed an auspicious start for the conference which by the late evening had drawn nearly 150 scientists from every corner of the planet.

Early the following morning, the full moon still bright over the blue-black Pacific, the breakfast bell tolled in the center of the conference compound, and soon the molecular biologists filed through the dawn light and into the redwood chapel that served as center for the next four days. Inside, the chapel was dim and gloomy, with theater-type seats, exposed beams, and an elevated stage that, even stripped of ecclesiastical accoutrements, was still unmistakably reminiscent of an altar. The implicit metaphor did not go unobserved. "Here we are," a young scientist from the East Coast told me later that night, "sitting in a chapel next to the ocean, huddled around a forbidden tree, trying to create some new commandments—and there's no goddamn Moses in sight."

Paul Berg was probably the jet-lagged congregation's closest candidate for the role of Moses, but that morning he appeared more like the quintessential young California academic: tanned, intense, trim, in casual sport clothes and a suitable collegiate sweater to ward off the early Monterey chill. He might easily have been dressed for sailing or an early round of golf, but instead he was standing in front of a curious and slightly tense audience of colleagues, expressing once again a concern that some in the select group privately considered obsessive. "What is new," he said with flat certainty, "is that recombinant DNA can now be made from organisms not usually joined by mating, and hence can give rise to DNA molecules not previously seen in nature."

"We can outdo evolution," was how David Baltimore, trimly bearded and clad in an embroidered Levi shirt, put it. At the outset, he labeled as peripheral the ethical issues of genetic engineering itself, as well as the potential use of the new techniques in biological warfare. While those questions are real ones, he said, they were not the specific charge of the Asilomar conference—a charge which the next four days would reveal to be plenty all by itself. "If we come out of here split and unhappy," Baltimore concluded, "we will have failed the mission before us."

And with that the mission began. Monday's first session, which rolled on well past the lunch bell, involved, appropriately, that old laboratory standby *Escherichia coli*—and soon grew oddly reminiscent of the ancient prophetic process of haruspicy, wherein seers predict the future by observing animal entrails.

The question, vociferously argued, regarded a variety of *E. coli* known as K-12, isolated decades earlier from

the stool of a diphtheria sufferer at Stanford, and by
now the bacterium of choice for much laboratory work
—including, of course, recombinant DNA. While *E. coli*
may well be the best-studied cellular organism on the
planet, there was one critical area about which no one,
clearly, knew much at all: How likely was it that this
E. coli K-12, so long laboratory-pampered, could survive in the human gut, should it escape its test tube
carrying some newly implanted and potentially unpleasant genetic information?

A series of British researchers proceeded that morning to demonstrate an experimental penchant for mixing
cultures of millions of K-12 into half-pints of milk,
swallowing same, and then monitoring their subsequent
stool for evidence of bacterial survival. The topic offered some opportunity for drollery ("A nice, quiet,
boring person," someone said as he interpreted a chart
of stool flora, "at least as far as his colon is concerned"), but by the end of the session, the implications
of K-12 ingestion seemed far from resolved.

While the English minimized the odds of the K-12
strain surviving in human beings, two American researchers, who had helped draft the plasmid group
report and who became central figures in the guideline-
writing process, were not so certain. Stanley Falkow,
a cheerful, unassuming microbiologist from Seattle,
pointed out that the *E. coli* K-12 is occasionally capable
of transmitting its plasmids, which would be carrying the
new genetic information, to other, related bacteria. And
that might be all it would take for the new information
to survive, in the gut, past the death of the K-12 strain
itself. And Roy Curtiss, an imposing, long-haired researcher from Alabama, suggested that thought should
also be given to the survival of *E. coli* in sewer systems,
where it might well have the opportunity to swap new
genetic information with literally billions of its fellows.
Falkow, for example, had just finished a mildly humorous presentation on the incredible number of *E. coli* that
could be harvested in the Potomac River, from the

sluggish waters right off the Kennedy Center in Washington, D.C.

These two speakers gave the first hints of what would begin to look like an inescapable fact: Before the advent of recombinant DNA techniques, human beings apparently knew more about *E. coli* than almost anyone would ever care to learn; but once we started manipulating that organism in ways not possible in nature, it grew distressingly clear that we really knew very little at all.

By noon on that first day at Asilomar, another controversy was taking shape, this one involving, in equal parts, the press, the public, and paranoia. It would raise problems that remain unresolved. "These proceedings," Paul Berg announced at the opening session, "will be taped, for the archives and for review, not for release. Anyone who does not want to be taped may ask and the machine will be turned off." Someone in the audience rose immediately. "But what about the *press*?"

There was a brief silence in the somnolent audience. What *about* the press, with their portable Sony cassette recorders perched stage left—all those $100 midgets right next to the sound equipment that cost the National Academy nearly $1000?

Howard Lewis, the NAS press officer, rose quickly and deftly reassured the scientists that all reporters use tape recorders, and that in fact good tapes contribute to a more accurate story. A vote was taken—the only vote of the first three days—and although there were many abstentions, a majority of those who voted were in favor of permitting members of the press to keep their recording equipment.

The controversy was hardly a surprise to the sixteen reporters in attendance. Since the first announcement of the Asilomar conference, press coverage had not been actively encouraged; at the outset it took some persistence for a reporter even to discover whom to ask for information. "Tearful," was how Berg described some of the early press requests.

That changed, however, when Howard Lewis—a

quiet, dark-haired fellow with an interest in science fiction—took charge. At that point Lewis was apparently one of the few organizers of the event to realize its potential historical significance. Besides opening up press coverage, he helped talk Berg into the official taping of the proceedings, a tape which by previous agreement will not be released for fifty years.

Along with the biologists and the butterflies, then, the press arrived—reporters from the major American and British science journals, the larger American newspapers, and a few magazine writers. But the atmosphere at the conference seemed hardly one of welcome.

"The scientists loved the press when we got Nixon," said the *Los Angeles Times* reporter. "But when we start hanging around their own backyard, they get very nervous."

"A secret international meeting of molecular biologists?" the *Washington Post* reporter claimed he'd told the conference organizers. "If the press isn't allowed, I'll guarantee you nightmare stories."

Some of the scientists, at that point, were clearly already having nightmares of their own. Stanley Cohen, for example—the brilliant young Stanford researcher who had isolated the plasmid pSC101 and whose initials were included in the name—was so press-shy at Asilomar that when the official photographer approached him at the chapel entrance, he retreated, face covered, like a newly arraigned mobster on the precinct-house steps.

Cohen is in fact a fairly agreeable fellow. But his reaction at Asilomar was hardly uncharacteristic of the conferees. Nor was it altogether unwarranted: A suitably hysterical story out of Asilomar regarding the antics of an international cabal of biologists devoted to some blackly humorous campaign of creating new cancer virus would have been just the thing to stampede Joe Public, attract the baleful gaze of legislators, and snarl molecular genetics in a skein of statutes for years.

But then of course, the more paranoid and cautious the scientists appeared, the more certain the journalists were that something was up. And their consequent behavior was not always reassuring.

After four days of intense sessions, in fact, some reporters—top names in their profession—were still asking questions that suggested they had spent the previous days locked in a very dark closet. As some cornered scientist explained for the fifth time a fairly fundamental concept of cell biology, the question in his eyes was clear: How the hell will this befuddled individual with the notebook be able to explain the subtleties of EcoR1?

Or perhaps even worse were those questions clearly designed to elicit the quotable lead for some mythical housewife or businessman to digest over morning coffee: "Dr. X, would you say that plasmid engineering is the most important advance in science since the mammal?" Or: "Dr. Y" (who has already expressed at least a dozen times the utter impossibility of answering this question), "in how many years will we have a cure for diabetes?"

"I don't enjoy talking down and trying to make things palatable for the breakfast-eating public," Berg would say later, "especially when it gets distorted and seems so mickey-mouse that I'm embarrassed by what it says."

Yet both during and after the conference, Berg was extremely patient with the press. And the coverage of Asilomar—which appeared in dozens of magazines and newspapers in the months following—turned out to be surprisingly good. Headlines ranged from "AND MAN CREATED RISKS" to "DECISION AT ASILOMAR," and accompanying editorials were almost always laudatory.

"Some of it," Berg said later, "had a more sensationalist flavor than others, but I don't think we could have gotten that quality of reporting if we had people standing outside the gate." Inherent in the example of Asilomar, from start to finish, was the question of just what position the press should take in the matter of scientific policy making.

"This is what we know how to do," one East Coast microbiologist plaintively told a reporter, midway in the proceedings. "This is what we're used to doing. I mean, we all get together, we want to know what everybody else is doing."

During the first two days of sessions, it was apparent that the conference attendees would really rather talk about almost anything but the issue at hand. "I felt," Maxine Singer said later, "that there was too much traditional science at the beginning." The program was laden with exceedingly technical papers, with titles like "Molecular biology of bacterial conjugation and conjugative mobilization of plasmid and other DNA's," or "Molecular cloning of DNA as a tool for the study of plasmid and eukaryotic biology."

Yet the field was sufficiently new and diverse that such traditional briefing was both welcome and necessary for many of the conferees. The talk was at least sufficiently original that after hearing certain papers, researchers queued up at Asilomar's two pay telephones to relay word back to their laboratories. The discussion helped define, moreover, just what was at stake in the experimenter's right to investigate the new technology. A major portion of it was in the realm of pure research—specifically, a process of cloning DNA fragments.

"Cloning" is a notion that has entered the public mind still bearing, unfairly, the ominous overtones it acquired in Huxley's *Brave New World*. A clone is, essentially, a cell or group of cells that has been derived from a single ancestral cell solely through the process of cell division, without sexual reproduction. It's simple to do with asexual organisms like bacteria: Put a single bacterial cell on a field of nutrient, take proper care, and fairly quickly it will divide sufficiently to be altogether surrounded by clones—identical copies of itself. As biologist François Jacob wrote, "The dream of a bacterium is to become two bacteria."

In higher organisms, it is less easy to do, and comparisons with Huxley's nightmare become more credible. But each cell of any given organism does in fact contain the total DNA coding for the entire creature, and both tobacco plants and frogs have by now been grown, with exceeding experimental care, from single cells.

Cloning of DNA, however, is quite different. It uses the recombinant techniques to insert a short piece of some creature's DNA into a plasmid, placing that plasmid into an *E. coli,* and then growing up a great mass of that *E. coli*—which also reproduces the plasmid. Ultimately one separates the plasmid DNA from the rest of the bacterial culture to yield something unprecedented in laboratory history: a concentrated quantity of a single DNA segment, each segment unimaginably complex in its coding, yet perfectly identical and ripe for intense biochemical analysis. The new access to large quantities of identical cloned DNA fragments was something that, by Asilomar, had biochemists' fingers twitching all over the planet.

It would be months before I absorbed the full implications of the new research capability implicit in recombinant DNA. At the time, about all I noted was how apt the phrase "building blocks of life" was for DNA; the schematic slides of various genomes looked just like elaborate Lego block constructions. The electron micrographs, on the other hand, looked like little more than earthworms in a mud puddle.

Two other facets of recombinant DNA technology made an immediate and lasting impression. The first was demonstrated when a tan young southern Californian clambered onto the chapel stage holding a three-foot tall weed, freshly harvested from a Monterey roadside that morning, as a colleague went down the aisle passing out similar plants to each row of biologists. The plant was a legume, the plant specialist announced, and if one shook the dirt off, it was possible to see the tiny bacteria-filled nodules on the roots that "fix" atmospheric nitrogen, pulling it directly from the soil and adapting it to forms useful to living organisms. That's not a bad trick, since most world food crops cannot do the same and thus require doses of artificial nitrogen fertilizer—usually created, in turn, from a commodity somewhat scarcer than dirt: petroleum. Clearly, if one were able to isolate the gene that teaches nitrogen fixation and then some-how ally that, say, with a wheat plant one would have an altogether remarkable food crop. And with the ad-vent of recombinant DNA, the plant specialist suggested it may not turn out to be such a difficult trick to teach after all.

The notion was clearly a more imminent prospect for the new style of genetic engineering than dial-a-baby; plants are already a venerable target of conventional genetic engineering, in the form of cross-breeding, to such an extent that there is currently considerable inter-est in establishing "gene banks" to retain the original forms of many now-hybridized plants, just in case some-thing should go drastically wrong with their updated ver-sions. Yet the whole notion of recombinant DNA in the plant kingdom went with only peripheral notice during most of Asilomar and the subsequent guidelines-writing process—an ironic indication, perhaps, of just how far genetics had traveled in the century since Mendel's sweet peas.

The second facet of recombinant DNA that was im-mediately visible involved commercial applications. The presence at Asilomar of representatives from three drug manufacturers—Merck, Roche, and G. D. Searle—along with a General Electric researcher, were reminders that recombinant DNA, even in those embryonic days, showed potential for providing remarkably cheap labor in the pharmaceutical industry. If it proved possible to transplant mammalian genes—say, the genes that code for insulin production—into an *E. coli* in such fashion that those genes will actually express themselves, that is, order the simple bacteria to turn out pure insulin, then one could support an entire insulin factory on a diet of sugars and salts. The insulin produced by means of that exceedingly delicate chemical template would almost cer-tainly be purer than the very best that could be produced with traditional techniques. And insulin might be only the beginning—the process could be applied to anti-biotics, growth hormones, a whole series of elaborate compounds whose production was either expensive or altogether impossible.

At the time of Asilomar, no one really knew how feasible it would be to program bacteria with genetic commands borrowed from higher organisms, although there were already schemes being thought up to circumvent whatever natural barriers might exist. The potential benefits of gene transplantation, even simply in the lower organisms, were clearly immense; what was also clear was that those commercial applications would likely involve large quantities of genetically altered bacteria and thus a concurrently increased risk.

But projecting future applications was not really the point of Asilomar. Some even suggested that it was a pursuit better left altogether alone; as soon as the pure researcher accepts credit for the positive applications of his discoveries, he must be equally willing to answer for its detrimental effects. And this, in the light of recent history, might not be such a bargain.

And at any rate, if something wasn't resolved about the moratorium itself, the state of the art was going to remain not only harmless, but also inert. Yet when, early Tuesday morning, the floor was opened for discussion on the first real matter of business—the lengthy recommendations of the plasmid group—there was utter silence in the redwood chapel.

The proposal was hardly uncontroversial. In thirty-five single-spaced pages, it established an elaborate six-stage classification of experiments by the degree of risk they implied, along with the safeguards required for performing each class of experiment, and even included one class of experiments that would be altogether forbidden. At first, there was not a single comment from the assembly. One of the five plasmid group members gave a brief pep talk, encouraging participation. Still nothing. Berg stared out at the audience briefly and then moved for the adoption of the report as it stood—and only then, with almost audible creaking, did the wheel of discussion begin to turn. And it proceeded to run right downhill into chaos. Odd, I thought at the time, that a roomful of the leading minds on the leading edge of science can't agree on how to run a meeting. But the proceedings that bright morning began to resemble some obscure primitive tribe, eons ago, accidentally stumbling by trial and error onto the secrets of parliamentary procedure.

There was, in fact, a great deal to talk about in the plasmid group report, which would provide the model upon which the next ten months of guideline formulation would be based. The six classes of experiments were carefully delineated. A class I experiment was one in which the potential biohazard seemed insignificant—the transfer of genes between similar strains of *E. coli*, for example. Class III upped the ante to the transfer of genes between species that normally could not exchange genetic information in nature—introducing random fragments of fruit fly or frog DNA into, say, *E. coli*, an abrogation of apparent natural barriers whose results, while likely harmless, could not be predicted. A class IV experiment involved the manipulation of genes that were known at the outset to be harmful. And class VI represented experiments so dangerous that they were banned —such as using *E. coli* to produce diphtheria or botulism toxins.

Each class of experiment would require a higher degree of physical containment, increasing the laboratory precautions against the culture being manipulated escaping—through exhaust fan, drain, or researcher's body— into the outside world. Those precautions, as we will see later, range from simply hanging a biohazard sign—a sort of sinister-looking mutation of the familiar radiation symbol—on one's laboratory door, to an elaborate set of space-age mechanical safe-guards that range from laminar-flow hoods to negative-pressure laboratories.

The containment levels, numbered also in the plasmid group's report, added up to a different kind of figure: the expense that each researcher would have to pay in order to upgrade his laboratory containment facilities sufficiently to allow the work he was interested in doing. At the time of Asilomar, Paul Berg had just finished modifying his own laboratory to a level that would allow him to do some, but hardly all, of the work described by the plasmid report. The price tag had run into five figures, and that kind of money was not a minor issue at Asilomar. Funding for research was never lavish. Who wanted to establish unrealistic containment requirements that might eat up research budgets even faster than Congress could whittle them down?

The first solid criticism that morning came from Joshua Lederberg, one of the two Nobel laureates in attendance. Lederberg expressed sympathy with the effort as a whole and then proceeded to voice an objection that, months later, would prove rather telling. "If we can't communicate the tentativeness of this document," he said, "then we are in trouble." With five minutes' consideration, he suggested, the plasmid group's paper appeared decent enough. But what final interpretation would it find? "There is a graver likelihood," he intoned ominously, "of this paper crystallizing into legislation than any of us would like to think."

Lederberg, a large, bearded, well-nourished man, wore loose, brightly patterned sports shirts as Asilomar, giving him the look of a senior California academic who spends his weekends in hot baths at Esalen or perhaps gardening. He had, however, two very different connections to the Asilomar proceedings. It had been his research into plasmid in the early 1950s, in the course of which he named those tiny rings of DNA, that earned him his Nobel prize. And a decade after that, Lederberg had been involved in what may yet turn out to be an extremely farsighted bit of scientific caution: convincing NASA that the biological isolation and decontamination of returning space vehicles was an absolute necessity, on the off chance that they might carry extraterrestrial infective agents against which life on earth would be defenseless. Lederberg's campaign resulted in containment facilities for spacecraft (and in several terrifying novels as well). It had, moreover, a curious parallel with the new issue of recombinant DNA: In both, the fear was speculative, the precautions costly, but the consequences of inaction were potentially disastrous.

"Nobel laureates can't believe in their own scientific fallibility," one young molecular biologist told me later that day, over lunch. "I've seen lots of them and it's common to the phenotype. If you're a Nobel laureate in this country, then there's nobody who can touch you."

That proved, however, not to be so at Asilomar. Lederberg, for one—a man of not inconsiderable social conscience—managed to emerge with the worst press of anyone in attendance. In part, it was due to the unrelenting obtuseness of his diction, and in part to the excessive attention generally accorded the Nobel laureates by the press. Lederberg ended up writing a letter to the *New York Times*, complaining about the misinterpretation of his position by a reporter for that newspaper.

In fairness, it was often almost impossible to figure out *what* Lederberg was talking about. Not so difficult to interpret, however, was Ephraim Anderson, the British expert on antibiotic resistance who had reacted so vitriolically to the misprinted version of the moratorium letter; at Asilomar, he was equally unsympathetic. After Lederberg's observation, Anderson rose immediately to ask the plasmid group a single question: "Which members have had experience with the handling and disposal of pathogenic microorganisms capable of causing epidemic *disease?*"

There was embarrassed silence, some laughter.

"It's no joking matter," Anderson persisted.

The entire panel finally sheepishly admitted that all of them had probably had a little.

Anderson—a portly, imposing gentleman—glared coldly. "If you're going to set down guidelines," he said, "you should *know* something of the matter. That's not a terribly profound principle; it doesn't take a great deal of cerebration to arrive at."

Anderson picked up the elaborate plasmid group report and read a selected sentence: " 'For our purposes, pathogenicity and virulence are defined similarly as "the ability to cause disease." ' " The Englishman closed the paper slowly. "This," he said, "must stand as the greatest oversimplification of all time."

He continued. The chairman of the group finally cut him off, thanked him for the "input," and asked for a written critique. "This is, after all," he apologized, "a rather terse document."

Anderson continued to stand. "You could have fooled me."

Alterations, the leader assured him, would be made; the working document was assembled, after all, in six days.

"And why couldn't you do it in six days?" the Englishman wanted to know. "After all, the Lord created the world in only seven."

The session lurched onward, sidetracking frequently only to end now and again in cul-de-sacs over extremely technical details: At what point in the laboratory manipulations can one consider a plasmid to be simply naked DNA? Would one really have to sterilize *everything* to work with *E. coli* K-12?

Before long, however, the discussion was rather forcefully brought back to fundamentals by the only other Nobel laureate in attendance, James Watson, then both Harvard professor and director of the Cold Spring Harbor laboratory. Watson, unlike Lederberg, seemed almost to cultivate the persona of the absent-minded academic: tall, pale, thin, shirt collar turned up, wispy brown hair tugged so constantly that it stood out from his head in total disarray. He spoke with a regular punctuation of grimaces, and in the midst of any given sentence, his gaze could wander off into space, a consummate 2000-yard stare.

Watson, of course, had been involved with the recombinant DNA question since the outset; his signature was on the moratorium letter. Thus it came as rather a surprise when, Tuesday morning, he rose to say that he thought the moratorium should end, and that it should end, moreover, without the complex kind of categorical restrictions that the plasmid paper represented. "Common sense," was Watson's prescription. "We can just suffer the possibility that someone will sue us for a million dollars if things don't work out." The proper response to the problem, Watson felt, was simply education.

Maxine Singer was on her feet immediately to ask precisely what had changed during the past six months to justify lifting the moratorium. Since, by definition, nothing *had* happened—at least experimentally—there was little Watson could say. "Some things are gross stupidity," he said, "but I don't know how we can legislate against it. We'll end up with situations in our labs where we know people are cheating a bit and ultimately we'll end up with massive dishonesty." Watson shrugged, looked around, sat down.

The discussion drooped noticeably. A young researcher from MIT—then working under Watson at Cold Spring Harbor—allowed as how he, for one, welcomed some guidelines. "There are experiments I would like to do," he said, "but I am not omniscient, and most of my experiments go wrong."

There was silence again. "We have to decide," someone pointed out, rather plaintively. "Is the moratorium over?"

Finally Sydney Brenner stood up to speak. Brenner is a compact Englishman in his forties, with bushy eyebrows, gleaming eyes, and a sort of nonstop animation, all of which blend to form an impression midway between leprechaun and gnome. Brenner, a late addition to the organizing committee, soon emerged as the single most forceful presence at Asilomar. Over the three days that remained, whenever the sessions wandered off into some technical morass that threatened to swamp the larger concerns, Brenner deftly put them back on the track.

"Does anyone in the audience believe," he asked Tuesday morning, "that this work—prokaryotes at least—can be done with absolutely no hazard?" He waited through a long silence. "This is not a conference to decide what's to be done in America next week," he continued finally. "If anyone thinks so, then this conference has not served its purpose."

There are two separate elements to the problem, Brenner explained. "Objective scientific things, which we can reach agreement on, and then an extraneous set of elements which might be called political. In some countries, this might be done by the government, and once the guidelines were set, and you broke them, there'd be no question about the censure of one's peers —the police would simply come out and arrest you." This is an opportunity, Brenner concluded, for scientists to demonstrate their ability for self-regulation—"to reject the attitude that we'll go along and pretend there's no biohazard and hope we can arrive at a compromise that won't affect my own small area, and I can get my tenure and grants and be appointed to the National Academy and all the other things that scientists seem to be interested in."

It was the first time Brenner had spoken out at the conference, and the effect was undeniably tonic. Equally tonic, in an altogether different way, were Brenner's activities following the noon lunch break. During the long hiatus between lunch and the next scheduled session, while more than a few conferees enjoyed the remarkable Asilomar beachfront, Brenner held an informal chapel meeting aimed at what in the months after developed into a curious new fillip in biological research: "disarming the bug," as Brenner referred to it—the deliberate creation of "ecologically disabled organisms."

"What I would like to do," Brenner announced at the outset of the afternoon meeting, "and what certainly seems incumbent upon me, is to erect the highest barriers possible between my laboratory, where the work is performed, and the people outside."

One set of barriers was obvious: physical containment, the use of well-established techniques for keeping any nasty new microbes safely inside the laboratories where they originate. A second notion, crystallized by the pressure of the moratorium and Asilomar, is something altogether new, and unprecedented in the history of human interference with nature: "self-destructing vectors."

In the language of mathematics, a vector is a line that possesses both magnitude and direction. In biology, it is strikingly similar: A vector, classically, is the carrier of some disease-causing agent. The vector for bubonic plague is the flea, for malaria the mosquito, for spotted fever the tick. In molecular genetics, however, a vector is the vehicle for an agent of very likely unknown consequence—the plasmid or bacteriophage that carries the foreign DNA fragment into the bacterial host.

The concept of biological containment, then, meant interfering with the process of natural selection in so negative a fashion as to produce plasmids, phage, and bacteria that are, by design, incapable of living in the real world, outside the laboratory. Thus even should they manage to escape into sewer or stomach, they—and whatever novel genetic material may have been engineered into them—would die without reproducing.

Brenner, much earlier, had jokingly suggested that the bacterial host for recombinant DNA work should be, rather than *E. coli*, *Pasteurella pestis*—the plague germ. It would be so dangerous, Brenner said, that everyone would be petrified to use it. "Either that, or we design the bug so that essentially nothing can go wrong."

And that was the purpose of Brenner's Tuesday afternoon brainstorming session. Forty or so were present, scattered through the first rows of the chapel, and the discussion—stratospherically rarified talk about "deletion mutants" and "auxotrophics"—rapidly grew heated, arcane, and confined to only the six or seven people who could keep up. The reporters filed out rather quickly, followed in turn by half of the scientists.

But the session itself was a considerable success and marked, in fact, the first of two major turning points in the Asilomar proceedings: By the end of that afternoon the odds on fairly rapid construction of the new, crippled strains looked so good that onlookers had renamed Brenner's "Mark One" safe vectors "Mach One."

That confidence was, to say the least, excessive, and one can only wonder whether Asilomar might have ended differently, had the real difficulties—and the long dry season before the return to the proscribed work—been fully known.

6

Asilomar II:
The Dilemma Intensifies

By late Tuesday afternoon, the early spring air at Asilomar was crisp and the sky still a cloudless Kodachrome blue. Inside the chapel, however, the curtains were drawn, the air was still and heavy, and only an occasional shaft of sunlight managed to penetrate, striking a balding head here, a graying one there. The texture and color in the rustic chapel were early Rembrandt; the content of the program, however, was pure and simple modern dilemma.

The discussion that afternoon involved the guidelines suggested by the group assigned to study the manipulation of DNA from animal viruses. The fear that oncogenic—tumor-producing—bits of DNA from such viruses might manage to propagate and then somehow operate in *E. coli* had been a major concern of the moratorium letter. The possibility was only theoretical, but it yielded perhaps the most terrifying scenario of all: a contagious form of cancer-causing bacteria.

The response of the virus group, thus, seemed oddly muted; it was a single terse page that brushed aside the question of safer vectors and prescribed containment procedures not very different from those already in use. The paper was odd in another way as well: Of the three working groups at Asilomar, the virus group was the only report to include a minority opinion. "Given the limited amount of information available at this time," wrote the dissenter, "I believe that the risks associated with the widespread, semicontained use of the procedure exceed the rewards from the information to be gained."

The lone voice belonged to Andrew M. Lewis, Jr., from the National Institute of Allergy and Infectious Disease (NIAID), who took the podium late Tuesday to describe his experience with an obscure class of viruses know as "non-defective adenovirus 2-SV40 hybrids." The viruses themselves may have been arcane, but the implications of his talk were far less so. Lewis —in his thirties, conservatively dressed, with an unmistakably serious mien—was the first person in the United States to be burdened with the distribution of a brand-new, laboratory-created—and potentially hazardous—tumor virus.

Simian Virus 40, as earlier mentioned, is to the fairly recent study of oncogenic animal viruses much as *E. coli* has been to the more venerable pursuit of bacterial genetics. SV40 is not yet, however, as demonstrably benign as *E. coli*—and that uncertainty was a major factor in Paul Berg's decision, two years before Asilomar, to call off his attempts to place portions of SV40 DNA into *E. coli.*

While at that point Berg was the only researcher exploring the exotic possibilities of artificial DNA recombination, he was hardly alone in his use of SV40. By then, in fact, SV40 had been poked and pummelled in virus laboratories for more than a decade. And thus it may have been inevitable that in 1969, a natural recombination between SV40 and another virus was reported —a recombination that was rare and remarkable in that it was nondefective and thus capable of continued infection and propagation.

While SV40 causes ominous changes in human cells in glass, it is not at all clear what, if anything, it does to human beings on the hoof. The virus that it happened to recombine with, however—adenovirus 2—is one of a common family found regularly in people, particularly in the region of the adenoids; it usually is either harmless or else causes a variety of coldlike symptoms.

By the spring of 1971, just about when Paul Berg

was abandoning his plans for the artificial recombination of SV40 DNA, five nondefective adenovirus 2-SV40 hybrids had been identified. The mix of DNA represented by the new hybrid viruses represented an unknown hazard to human beings—and also an extremely interesting subject for research. Thus did Andrew Lewis, associated with the Laboratory of Viral Diseases at NIAID, find himself charged with the distribution of virus strains, of unknown hazard, to other research laboratories.

"The question we faced," Lewis describe it at Asilomar, "was whether one individual had the right to decide to distribute potentially hazardous laboratory-created recombinants." The decision was forced rather quickly, when the provocative new hybrids were described at the Cold Spring Harbor Tumor Virus Workshop in August 1971. The reaction to Lewis's reluctance was immediate: threats of congressional action, administrative pressure from NIH, even a group letter to *Science* should Lewis try to withhold the new hybrids. And on the other hand, concerned researchers warned that if Lewis went ahead with the distribution, they would file for a federal environmental impact statement.

"I felt," said Lewis, "that voluntary compliance by interested investigators was the most satisfactory method," and so he decided to require a formal document—felicitously dubbed a Memorandum of Understanding and Agreement—from each laboratory that requested the viruses, stipulating that the researchers themselves would assume full moral and legal responsibility for the novel viral agents.

About two years after that, Lewis stopped growing the virus altogether. And by the time of the conference at Asilomar, he was no longer so certain about the notion of voluntary compliance. "Several major laboratories," he said, "have thus far not supported the memorandum. In addition, our original request to restrict the distribution of the first hybrid seems to have been ignored by one or more of these same laboratories.

"The unwillingness of laboratory directors and interested investigators," Lewis concluded, "to respond to well-founded concerns and to accept responsibility for containing potentially hazardous agents, would appear to have significant implications for any attempts to deal with the problems posed by bacterial plasmid recombination."

But Lewis's implications, clearly, were not popular ones. The chapel audience remained quiet and cold, and midway through his presentation Lewis began to lean intently over the big wooden podium, grasping his pointer, straight up, like a spear carrier in an Italian opera. The source of the hostility seemed no mystery: Asilomar was, after all, *about* self-regulation—and Lewis's first-hand testimony was not exactly in the spirit of things.

By Wednesday morning, the situation at Asilomar seemed as unsettled as the Monterey weather. The mild blue skies had begun to turn and by afternoon a massive bank of thick gray fog hung off the coast, sending in low damp clouds both morning and evening. The conference meals had begun to deteriorate as well, and Wednesday's corned beef and cabbage had barely achieved summer camp standards. Worst of all, however, was that the conference, enmeshed in bickering, might actually not be able to arrive at a group statement. But clearly, with three days of talk about biohazards thoroughly soaked up by the press, it was too late to stop.

Odd communications from the outside world had begun to appear. A photocopied telegram was delivered to each attendee, sent by a Russian molecular biologist who had been denied permission to emigrate to Israel on the grounds that his work was "considered important for the state security of the U.S.S.R." The Russian hoped, with a kind of blissful desperation ("However crazy this may seem to you . . .") for a statement from the Asilomar group on the inapplicability of his work with recombinant DNA to biological warfare. No one talked much about it: least of all the three Russian delegates, the eldest of whom—a white-haired honcho in the Soviet scientific establishment—dismissed it with a shrug. "First he started sending these to kings and presidents," he observed mysteriously, "and now he is sending them to the porters at the door."

Another letter, mailed separately to each of the conference participants, came from a Cambridge, Massachusetts, group of radical scientists called Science for the People. The letter requested a series of massive changes in the whole decision-making process that, considering the chaos that already existed, seemed distinctly optimistic. "Science for the people?" blithely responded one researcher. "That's nice, but the only science I've done has involved pigs."

Tuesday night, the virus group report had developed into a vigorous, sprawling debate that managed to make the plasmid group discussion seem rather mild. And the virus argument was in turn soon to be dwarfed by the disagreements over the third and final working group report, presented Wednesday afternoon, on the little-examined prospect of cloning segments of DNA from the higher organisms.

The technical name for this subject was "cloned eukaryotic DNA"—referring to genes from all organisms higher on the scale of things than bacteria, from yeast through plants to human beings. The area was likely to be the most promising of all recombinant DNA technology and also the least understood.

One facet of the new technology bore an ominous handle indeed: "shotgunning." In a shotgun experiment, one isolated the total DNA content of a cell—the sum of its genetic information—and then chemically blasted it into individual, short fragments using one of the restriction enzymes. Each resulting fragment, depending on its exact length, then contained as few as two or three discrete genes, each coding for one or another kind of enzyme or protein production.

Each fragment, however, also retained the characteristic "sticky ends," and could thus be functionally reincorporated into something like a plasmid. By growing cultures of those modified plasmids, one could end up with a labful of test tubes, each containing some distilled, amplified, and chemically active portion of an organism's total genome, thus providing the basis for a kind of hands-on mapping technique light years past anything Thomas Hunt Morgan likely ever dreamed of with his fruit flies.

The isolation and identification of mammalian genes is a critical step in a whole series of projected applications for molecular genetics, from providing the genetic blueprints for those bacterial drug factories, to, in some distant future, gene therapy, wherein a malfunctioning gene that causes, say, some hereditary disease in a child, might be replaced or backed up by a healthy one.

That will likely be in the distant future indeed. Frogs, yeast, and fruit flies were the sole subjects of the first tentative shotgun experiments at Asilomar. But the potential risks were already apparent.

One of those risks had been mentioned in the moratorium letter. Certain segments of mammalian DNA, it seemed, appeared strikingly similar to the structures of tumor viruses. The letter warned that liberating those sequences with recombinant techniques might result in an experiment that started out with harmless animal DNA and ended up with a tumor virus, perhaps safely ensconced within an *E. coli* bacillum. While the nature of those "cryptic viruses" was speculative, theory had it that they might normally be kept in check by other genes further down the DNA molecule. Removed from that genetic context, however, it was impossible to predict how they might function.

It was a chilling indication of just how unknown the new territory was. The Asilomar working paper on eukaryotic DNA went even further: Since no one knew the individual nature of the genes being shotgunned, it might even be possible to introduce accidentally some foreign gene into a bacterium that, by chance, happened to code either for toxin production or else fundamentally alter the host range of the bacterium itself.

The concerns were exceedingly speculative, contingent on everything from unproven notions about cancer to the equally unproven potential of mammalian genes to operate inside bacterial hosts. But once again the potential consequences seemed sufficiently grave to demand extra caution.

The working group on eukaryotic DNA had produced a paper midway in scope between the plasmid and virus group efforts. To head the group, Berg had selected a Carnegie Institute researcher named Don Brown—a rangy, crewcut, mildly abrasive man who had been one of the first to use the new recombinant DNA technique. His object of study had been the African frog *Xenopus*, a small green amphibian whose agreeable fecundity has long made it a laboratory favorite. Brown's views on the safety of cloning eukaryotic DNA were thus, by familiarity, rather liberal, but Berg considered Brown open-minded, and thought that if Brown himself could be persuaded on the matter, then the recommendation itself would carry that much more weight.

The notion would, however, once more underscore the unavoidable conflicts of interest in even the most benign use of insiders, when, late Wednesday afternoon, the eukaryote group members took their seats on the chapel stage, and presented the eukaryote paper to the assembled.

Their recommendations followed the plasmid group's lead of numbered categories of risk, and that, right away, launched disagreement.

Joshua Lederberg, still worried about keeping things tentative, but giving a bit of ground, suggested that the specific categories might be reduced to something like high-medium-low.

"But shouldn't we benefit," someone asked, "from the experience we have?"

James Watson, slumped low in the middle of the audience, muttered to his neighbor: "But there *is* no experience."

"I have to emphasize," said Brown, on stage, "that there was a great deal of consensus among the members of our panel."

"So is there in the State Department!" Watson exclaimed softly. He sat for a moment and then whispered quietly to his seatmate, "There people have made up guidelines that don't apply to their own experiments."

"Stand up and say it," his companion urged softly. "You can say it; I can't."

Finally, with sufficient prompting, Watson rose to ask, "Why, according to the panel, is *Xenopus* DNA safer to work with than, say, cow DNA?"

Chairman Brown frowned, and looked slightly embarrassed. "It wouldn't be fair for me to answer that question," he said, and turned to the panel. "Anybody like to defend *Xenopus?*"

While there were, in fact, some defenses to be made, no one at that point felt like doing it. Finally Watson sat, shaking his head. "He refused to answer the question," Watson announced softly to anyone within range.

Paul Berg stood to get the session back on course. "We have to make a decision," he said. "Can we measure the risks numerically?"

Watson, *sotto voce*, exploded, "We can't even *measure* the fucking risks!"

From there on, the discussion began to fragment. Roy Curtiss, the long-haired researcher from Alabama, suggested sensibly that they should make sure that "anything that comes out of this meeting should self-destruct in twelve months." A Stanford worker wondered what would happen when local committees met to assess specific biohazards. "If we can't agree on the danger of experiments here, imagine the situation of a local university committee!"

"Legislation," said one experimenter, "is inevitable. I can't believe that we'll be allowed to continue to control ourselves. But something worse than legislation would be if, in a few years, there's suddenly an epidemic around Stanford, say, or Cold Spring Harbor."

Just before the dinner bell rang, Anderson stood to suggest that the problem was clearly sufficiently complex that those in attendance should go home, brood a bit, and then offer suggestions in writing.

"So you don't believe we can arrive at a statement by Thursday noon?" Berg asked him.

Anderson paused, shook his head, displaying his first public expression of sympathy. "I don't know," he said finally. "I don't know."

"But we are the people," said a young American, "who are supposed to *know* about this and we can't go home from here without deciding anything."

"We haven't," Lederberg said, "been told what the *vector* will be. Unless Berg and the organizers tell us the utilization of this document, I'm going to be very hesitant about making any recommendations."

Berg stood again. "If our recommendations," he said "look self-serving, then we run the risk of having standards imposed. We must start high and work down. We can't say that a hundred and fifty scientists spent four days at Asilomar and all of them agreed that there was a hazard—and they still couldn't come up with a single suggestion. That's telling the government to do it for us."

At this, Watson, inspired, was up like a shot. "We can tell them they couldn't do it either!"

Barely audible beneath the laughter was Berg, again. "But we can't just let this thing drift . . ."

But drifting, however, was precisely what the conference seemed to be doing. By early Wednesday evening there was hardly a sense of anything approaching unanimity, and it was difficult, moreover, to see how one might develop. But some thing was already happening beneath the surface of the conference and, by chance, a new impetus was just about to arrive.

The Wednesday night program looked fairly innocuous: the presentations by the lawyers recruited by Daniel Singer regarding ethics and legal liability. It seemed a relief—lawyers are at least supposed to have some knack for public speaking and the evening promised, if nothing else, a bit of diversion.

And so, at first, it seemed. Daniel Singer led off—dapper and goateed, in black turtleneck and wire-rim glasses—and spent a mild-mannered quarter hour dissecting the dictum familiar to all physicians—"first of all, do no harm"—and concluded with a thorough three-part breakdown of risk-benefit analysis that neatly subsumed the major issues of the previous three days. "We should not pretend," Singer concluded, "that we are not making ethical judgments."

So far so good. While Singer's analysis wasn't going to help anybody decide between numerical risk ratings and high-moderate-low, it seemed something of an outsider's benediction: a statement of how important the conference's responsibility was, and just how complicated the problem.

The second speaker was Alex Capron, a professor of international law at the University of Pennsylvania. Clad in a sports coat and an open-collar shirt, he looked hardly past legal drinking age himself, and began with a nice joke: A scientist and a lawyer are arguing about which of theirs is the older profession. The argument goes back and forth, from Pericles to Hippocrates to Maimonides to Hammurabi, until it reaches all the way back to God. God, the scientist states, must have been a scientist, to have brought order out of chaos. Yes, the lawyer responds. But where do you think the chaos came from?

The joke turned out to be less joke than promise, as Capron, after a brief appreciation, launched into a merciless "outsider's analysis" that within moments had jaws dropping all over the chapel. Much of the previous discussion, he suggested, had been altogether irrelevant to the central issue.

The Asilomar audience grew suddenly very quiet. Many of the arguments advanced against establishing guidelines, Capron said, had been equally inapplicable. "Academic freedom," he noted acidly, "does not include the freedom to do physical harm." And "prior restraint" —a notion that Lederberg had briefly floated on Tuesday—makes perfect sense and is justified when it involves restraint from doing physical damage.

"This group," Capron suggested flatly, "is not competent to assign overall risk."

The question of earlier that day instantly came to mind: If *we* can't do it, who can?

"It is the right of the public," Capron continued confidently, "to act through the legislature and to make erroneous decisions."

That was hardly reassuring to hear from a lawyer, and it was clear that the audience in the redwood chapel was growing a bit uncomfortable.

Capron then made it worse, by suggesting a hypothetical situation wherein Congress itself might insert its grubby political fingers into the delicate process. Capron backed off, however, admitting that he thought legislation would be unnecessary—but even so, the very mention of "Congress" had drawn a low but audible groan.

But then again, the lawyer continued, legislation might not be all that bad: The law might provide, say, liability insurance for biohazard accidents.

That seemed faint comfort. "Legal institutions," Capron intoned in civics-class fashion, "are a part of your world, whether you like it or not." And considering the implications of recombinant DNA engineering, Capron concluded, it was time to involve those institutions.

A low buzz of conversation filled the Asilomar chapel while the next speaker was introduced: lawyers' night was clearly developing into something less pleasant than technical papers. One might well have viewed it as one group of arrogant professionals being dressed down by another group of at least equally arrogant professionals. And the lawyers probably made more money, too.

But the best—or perhaps worst—was yet to come. The final speaker was a lawyer named Roger Dworkin —short, slight, middle-aged, fairly nondescript, wearing mismatched suit and tie, thick glasses. One would have noticed him earlier only because of his ceaseless squirming during the technical sessions; the scientists did not squirm—they either paid attention and took notes or went to sleep. Dworkin looked like something of a manic milquetoast, but by the time he had uttered a sentence or two, his sharp, dry delivery had made it clear that here was the sort of lawyer who slices one mercilessly into very tiny ribbons in the witness box. The necessity of appearing in court was, coincidentally enough, precisely his topic: "Conventional aspects of the law," as he put it, "and how they may sneak up on you —in the form, say, of a multimillion dollar lawsuit."

Abruptly, the audience grew very quiet. The subject of legal responsibility had not yet been dealt with in the fundamental terms of precisely who gets sued should something go dreadfully wrong.

Dworkin's job was to do just that, and having himself just squirmed through three days of the scientists' abstruse technical jargon, he took some relish in trotting out some of his own profession's—torts, liability, proximate cause, OSHA—to illustrate just how finely, if not fairly, the wheels of law can grind.

Professional negligence, Dworkin announced, is a failing that juries are exceedingly unsympathetic to. He cited the case of an Oregon ophthalmologist who lost a malpractice suit on his failure to perform a glaucoma test on a young and asymptomatic patient in whom the chances for glaucoma were one in 25,000. Judges, he pointed out, are experts only in law—and juries, quite intentionally, are experts in nothing at all.

It's not, however, Dworkin reassured, totally hopeless. If, say, a burglar were to break into one's lab and steal a vial of deadly virus and then strew its contents all over Brooklyn, then maybe—just maybe—the intervention of a third party might get one off the hook. Or if one's work was for national security purposes, then one might also be fairly immune from prosecution. But professional negligence itself is a broad concept: It can mean either performing an act poorly—or *performing it at all*. Thus one could even follow the guidelines, said Dworkin, and still be perfectly fair game for a lawsuit.

Past that, there may already be laws on the books that could apply to recombinant DNA engineering, he suggested. OSHA, for example—the Occupational Safety and Health Act—could conceivably be invoked to protect laboratory workers. According to OSHA, the lawyer explained, "the workplace must be free of hazard. Not relatively free," he emphasized. "The statute says *free.*" And the person who sets those standards is the Secretary of Labor.

Months later, an OSHA representative in fact attended another recombinant DNA guidelines meeting in California, listening for a day and then drifting back out —explaining, quite reasonably, that protecting a few hundred laboratory workers from a fairly hypothetical hazard would for the moment have to take a back seat to the difficulties of, say, protecting pregnant women who work in steel mills. Yet at Asilomar, the very notion of OSHA managed to provide the most violent intrusion of the real world into the proceedings thus far—a vision of some civil servant from the Department of Labor waltzing into one's lab for a surprise inspection on the outcome of which hung a $10,000 fine.

Or, for that matter, the notion of one's own laboratory technician, bizarrely diseased and setting out for revenge on the basis of a bloodless legal principle that Dworkin blithely termed "deepest pockets." While there had been no lack of real and humane concern during the first three days of Asilomar, there was something about the brief legal seminar that brought home forcefully just how *unpleasant* things could actually get.

Moments after Dworkin concluded his presentation and the lawyers took their seats on stage, the molecular geneticists rallied to the defense. The more vocal bravely cited legal precedents—fetal experimentation, medical research on prisoners—as confidently as they were able. The effort, however, was just about as effective as if one of the lawyers had risen earlier to question the precise details of a certain enzymatic manipulation of bacteriophage lambda.

Joshua Lederberg finally took the lead and with some eloquence elaborated an intricate analogy involving the risks and responsibilities of accidentally importing a deadly African virus.

"That argument," said Dworkin, "with all due respect, is almost entirely beside the point. If we are remiss about our international travel regulations, we should move to correct that situation, rather than taking it as reason for being equally remiss about our approach to the biohazard question."

Lederberg returned to his seat, big tan arms folded across his chest like a wounded Buddha.

"This body," Dworkin continued, "probably doesn't even know its power. The law has a tradition of listening to and respecting expert groups that regulate themselves. On the other hand, there is precedent for ruining groups that don't—physicians, for example. Malpractice law has always been skewed in the direction of the physician, physicians have refused to testify against each other, and as a result they are now massacred in court."

"You misstate my case," said Lederberg, rising, and then a groan went through the chapel.

"Other people would like to talk too," someone said at the back of the room.

"Sorry," Lederberg mumbled, and sat again quickly.

The next question was more along the lines of *precisely* who is likely to get sued? And by the end of the evening, one of the lawyers was actually advising the conference to look into the possibility of extended liability insurance. "At least," he said, "then you won't have *quite* so much to worry about."

7

Asilomar III:
Final Hours

Thursday provided the grayest dawn of all for the International Conference on Recombinant DNA Molecules. And most of the organizing committee were still awake to see it. The night before, commencing just after the evening session with the lawyers, had been devoted to a nonstop round of manic writing and rewriting to draft some sort of coherent statement that the conference, as a body, might adopt.

At the time, it appeared almost impossible. Each evening, following the formal sessions, a small, dimly lit Asilomar conference room was stocked with tubs of bottled beer, and each evening the room was packed by ten o'clock. "Wednesday evening, at the beer session," Berg would say later, "you could see that people recognized that if we came out with nothing, it would be a disaster. And there was almost a feeling of helplessness, because nobody knew how to come up with what was needed.

"But I don't think," he added, "that we understood how important the beer sessions were. That's where all of the really intensive interplay of people's feelings came out, where positions changed and where people who had come with one view changed their minds."

Wednesday night, however, the organizing committee had no real idea whether minds had changed or not. "As the meeting was going on," Berg said, "I was worried because I could not pick out a common thread. I could not see the consensus. People, I think, were being very self-serving. As I'd discovered two years earlier, everybody would like to draw a circle around their work and stamp it as pure and unadulterated—it's what *you're* doing that is nasty and needs to be proscribed."

"Even if they voted our statement down," Maxine Singer said later, "we'd agreed to send it in ourselves." The statement materialized early Thursday morning as a freshly duplicated five page handout titled, "Statement of the Conference Proceedings."

The first paragraph set the tone: "The new techniques combining genetic information from very different organisms place us in an area of biology with many unknowns. It is this ignorance that has compelled us to conclude that it would be wise to exercise the utmost caution. Nevertheless, the work should proceed but with appropriate safeguards."

The Statement was, clearly, a compromise: "The six-category classification of risk was now uniformly condensed to high-moderate-low. The virus group recommendations had been tightened. But there was still no flat proscription of the experiments some had earlier called unreasonably hazardous.

Yet it was nonetheless a strong statement: If adopted, many researchers would have to go home and spend thousands of dollars on new laboratory containment equipment in order to do the experiments they could have done at no extra expense eight months earlier. The notion of comprehensive new safety regulations could hardly have been comforting. "Already," one researcher told me that morning, "we spend two months a year applying for grants; now we're afraid we'll spend another month filling out more forms. And the forms don't protect anybody; they just take more time."

The statement was handed out as scientists walked into the chapel. At nine, Berg stood briefly to announce a half-hour reading period while the organizing committee rather fitfully occupied their chairs on the stage. Their uncertainty became clearly manifest at the end of the reading period, in a short-lived but valiant effort to keep anything from actually coming to a vote.

The attempt started briskly enough. Berg concluded the reading period with a crisp observation: "I would like," he said, "to terminate this meeting at noon, and I hope that by then we will have reached the point at which that is possible." Because of that deadline, Berg suggested, it would be best to reach consensus by means other than voting. "This is not a statement of the conference," he reassured the audience, "it is a statement from the organizing committee, an attempt to pull together the views as we see them."

The statement was, however, not *signed* by the committee, and while Berg hastened to add that the omission was an oversight, it didn't exactly get the proceedings off on the very best foot.

The trouble began with the first paragraph. A minor quibble over wording arose and Berg reassured the researchers that all of the material discussed that morning would be included in the final report. "I take your comment to heart," Berg said, "and I'm sure we intended it, but somehow it got left out."

Stanley Cohen, from Stanford, rose immediately: "Do I understand that last comment to mean that there will be no opportunity at this point to modify the report that is issued—but that comments will be incorporated in some later, final report?"

"Yes," said Berg, "that is the correct understanding. This statement represents our assessment of the consensus as it exists."

"Well," said Cohen, "I for one would take issue with that." The report, he pointed out, was written by only five people. "We're wasting our time here this morning if we accept this as a final document, and the input of a hundred and fifty people over the next three hours is ignored." He shrugged. "I don't know if anyone else here shares that opinion, but it's mine."

David Baltimore, on stage, rubbed his eyes wearily. "I see no reason why you should believe that your input will be ignored. If we wanted to ignore it, we would be asleep right now."

Cohen was not satisfied. "I'll sit down," he said, "but I just want to say that I don't see it as simply as that, David. The report that comes out of this meeting will be seen as the report of a hundred and fifty scientists who came together to deal with this important question."

"This wasn't," Maxine Singer said, "written in a vacuum." Berg tried to follow up, reassuringly, by describing the document as a "working paper."

"Do you have any idea as to the mechanism for determining this representative consensus?" someone asked from the audience. "Are you talking about a vote on each paragraph?"

"Nope," Berg said firmly. "No. We're not talking about a vote. I think we'll get an idea about consesus in some way without a vote."

The word "Provisional" was quickly added to the title of the statement. But even that was not enough. David Botstein of MIT stood to say, rather plaintively, "Somehow—maybe we can tell the press to go away—we should get some feeling of whether eighty percent of the people in this place like it, or hate it, or *something*. Because I don't want to shuffle off my responsibility on you. I think maybe an overwhelming majority of the people here are willing to commit themselves to something of this sort."

Botstein had a more accurate sense of the feeling than did the organizing committee at that point. Berg tried once more to sidestep the vote: "David, I think perhaps there'll be some mechanism for achieving what you've suggested, in a more informal way but what I'd like to do is return the discussion to the principles in the document and find out whether a consensus does or does not exist."

By then, however, Sydney Brenner was signaling for attention, and it was Brenner, finally, who faced what was clearly inevitable. "I think," he said slowly, "that the first paragraph contains an extremely important statement of principle, and I would in fact be quite willing to see how that is received by showing hands." And after a brief summation—work should go forward but with sigficant controls—Brenner called for a vote: "And so who is in favor of it?"

The vote was virtually unanimous. Berg, looking slightly puzzled, resumed his chairman's duties. "And are there any opposed?" No hands appeared at all.

"After that first vote," Maxine Singer said later, "I relaxed, because the vote was so overwhelmingly different than what I expected that I realized then that I was a lousy politician."

"It was then," Berg agreed, "that we realized that we'd been listening to the wrong people. A few people were doing all the talking, and a lot of people had been quiet. And the quiet ones were in favor of coming out with something just as we had, and it was the Lederbergs and the Watsons and a few others who were doing all the talking and confusing us. We thought they were reflecting what everybody wanted and felt."

Apparently not. The 150 scientists at Asilomar were almost unanimously in favor of regulating the exotic new technology. Precisely how to regulate it, however, was another matter altogether.

At first the objections were minor. "In the middle of the first paragraph of section two," Ephraim Anderson intoned, "the preposition 'to' should be 'with.'"

Berg eyed him coldly. "We could have used you at four o'clock this morning."

Anderson smiled. "You can *always* use me."

There was brief laughter and then Anderson followed with more trivial complaints, but he was soon interrupted by David Baltimore. "We have," said Baltimore firmly, "very important substantive matters to discuss here this morning. If we worry about our grammar, if we worry about the details, then everybody's going to have their own opinion. We freely admit that it's not the best we can do. We'll try our best to do better, and we'll send it to Dr. Anderson to check the prepositions, but I'd prefer that the very valuable time we have now not be taken up with such questions. Please."

Anderson rose, clearly wounded. "I'm sorry," he said, "but that prepositional nicety involved a matter of semantics . . ."

"Yes, yes," said Berg quickly, "we accept your correction as it stands . . ." and his reply trailed off into assembled laughter.

The laughter soon yielded to the first major disagreements, over the notion of "most-dangerous experiments"—the idea advanced initially by the plasmid group report that there were some manipulations of recombinant DNA that simply should not be attempted under present conditions. But there was no real mention of excessively dangerous work in the final report, and one long-haired California researcher, in baggy sweater and thick eyeglasses, suggested rather vociferously that there should be. "What I would like to see," he said, "is the statement that *some* of the work should go ahead."

"It's not spelled out," Berg agreed.

"We could insert," Brenner suggested, "that there were some members of the meeting who felt that there are experiments so dangerous that they cannot be safely executed with any currently available containment procedures. I believe that is a statement of fact."

There was a brief silence. "Is that 'some members,'" asked someone, "or the majority?"

The question was hardly an idle one. If there was any element of philosophical discomfort in the noble self-regulation attempt, it was in the notion of the right of free inquiry. The history of science records myriad abridgments of that right, involving individuals like Copernicus as well as whole generations of scientists like the Soviet biologists trained under the eccentric genetic doctrine of Trofim Lysenko. There are sufficient examples, certainly, to make most modern researchers a bit paranoid about even the most apparently benign and humanitarian efforts to control the boundaries of their own curiosity. And while the whole notion of the moratorium may have been interpreted by some as a similar abridgment of that right, nowhere was it more sharply reflected than in the idea of flatly proscribed experiments.

"Well," said Brenner, a bit impatiently, "let's get the feeling. Is it *some*, or *most*, or *all*?"

"I would imagine," said one American researcher, "that it would be politically and intellectually sound to say that 'many' of the experiments should go forward, rather than all of them."

Berg called an immediate vote, and the notion of 'reservations' passed almost unanimously, and for the moment, the issue slid to one side.

From the mildly philosophical the discussion promptly plunged into the minutely technical as the meeting began to consider the specific containment recommendations, and precisely how one should do one's duty in terms of keeping potentially hazardous microorganisms in the lab. This, without doubt, was where the deep water ran, because while there was really no doubt among the assembly that recombinant DNA technology, as a whole, represented some kind of potential hazard, there were more than a few opinions as to exactly where that hazard resided. And those opinions clearly depended on precisely what field one happened to work in.

Berg did his best to prod things past this sticky juncture, but the morass of disagreement was deep indeed. At the outset of the quibbling, Sydney Brenner interrupted from his stage-left seat once again: "I would just like to say," he said wearily, in almost a half-speed caricature of his preciously animated delivery, "that we have no legal power here to compel *anybody* to do *anything*, and this has been so since the first letter.

"The letter asked people to take this into consideration. It asked people to defer. On the whole, it was widely accepted. Now—what we are trying to do *here* is decide whether we should further slow up, make further considerations, before we rush pell-mell into the field. I don't want, and I've said this over and over again, to carry the can for *anybody* here. If we make a definition such that by some rewording it is thought by some people that this allows them to try certain things—and let us just say something *happens*—then I don't want those people to say, well, I did it because the Berg committee said I could. I don't want to carry the can for *any* of you.

"You may well be overtaken by events which are out of our control," Brenner continued. "We don't know what the future is. So I think all we can say here is that, in our consideration, certain experiments should have to cross higher barriers of judgment, and others will have to cross even *higher* barriers of judgment. That is all we can accomplish here. We are not trying to cover all posble situations. We have no power, other than moral censure, if someone goes off to Uganda tomorrow and puts General Amin into plasmids."

Berg used the laughter to move the meeting forward— but not for long. It quickly became clear that if there was any rock upon which the notion of Asilomar would founder, it would be the rock of specific recommendations.

Joshua Lederberg, who had earlier railed at the excessive specificity of the six-level classification of risk, had by now changed his mind. "The gap," he said, "between 'low' and 'moderate' is an enormous one from an operational standpoint. The plasmid report suggested a number of intervening stages which were entirely reasonable. But the difference between low and moderate is considerable reconstruction of a laboratory facility. I do not accept it as a reasonable statement of procedure, because it is too severely step-functioned."

Someone else followed up immediately, suggesting that it looked, upon close reading, as if certain kinds of research allowed under the moratorium letter were now to be deferred until a safer vector was developed. And thus the Asilomar conference would not in fact end the moratorium, but *extend* it. Was that the intention of the committee?

It was quite true. No safe vectors then existed, and much of the research depended upon their availability. Richard Novick hemmed and hawed, but finally admitted, "It was the judgment of the committee that when one reaches the level of the warm-blooded animals, the risk of encountering tumor viruses was great enough that one didn't want to fool around without built-in containment, both physical and biological."

"Okay," Berg said, "I think that's enough. We'll go on to describing the three kinds of experiments—low, moderate and high."

And for the outsider, this was doubtless the most dizzying and impressionistic hour at Asilomar: an episode of technological horse-trading conducted in a vocabulary that often seemed to fit the realm of science fiction rather better than that of hard science. Because the fact remained that no one, at that point, really knew what these new manipulations of DNA might produce.

And thus no one really knew any more about the inherent hazards than they had at the outset. "Nature," as one researcher had said earlier, "is our only experiment." And whether nature had in fact tested and discarded myriad recombinations of unlikely DNA's, or had managed hardly any at all, was completely unknown. Nature keeps scanty records and, moreover, is never held accountable. But the researchers would be. And so the specific decisions were, finally, unavoidable.

At first there were vaguely parliamentary attempts to insert specific rewordings into the statement, but at the beginning of particularly labyrinthine suggestion about antibiotic, resistance, Berg intervened. "What I would like to suggest," he said, "is that we receive these recommendations in writing, and we will redraft the statement and have it sent to you, and that way can all have a version that reflects these comments."

Berg soon announced, rather optimistically, that he sensed a broad consensus in support of the section that dealt with introducing novel genetic material into bacteria. But one issue remained unresolved.

"One of the principles that was included in the prokaryote statement," said the long-haired California researcher, "was that there are certain experiments that should not be done at the present time. I notice that this particular series of paragraphs has left that principle out."

"May I respond to that?" David Baltimore asked. "That was a conscious decision, based on the demonstrable split of opinion about whether this is a philosophical question regarding right of free inquiry, or a question of science. Our decision, which is certainly reversible, was to put such experiments in the high-risk category, so that such work would require high-containment facilities. That is by no means a license for anybody to undertake lightly anything they get the idea to do, since those facilities are both extremely cumbersome to work with and not widely available. If there is significant disagreement with that principle, then we should discuss it right now."

"May I emphasize," said the young Californian, "that it is a serious omission not to acknowledge that the scientific community is split on this issue?"

"I think," said Berg, "that we can incorporate that view into our document." The concession would not be sufficient to jettison the issue entirely, but at that point Joshua Lederberg was again asking for the floor. "Josh," said Berg, "one last remark on this and then we'll move on."

"Well," said Lederberg, "I think the best way to understand general principles is to ask how they're going to apply to specific cases. So it will save me a great deal of trouble later to get your judgment on a specific case. Under the doctrine Dr. Novick has just recited, would it require a moderate risk facility, comparable to that used for oncoviruses, to introduce pSC101 into *Bacillus subtilis*?"

There was along silence. Lederberg had chosen a difficult specific: Even the original, far lengthier plasmid group paper hadn't been definitive about the kind of work he'd described, and in months to come, in fact, the hazard rating would drop. But for the moment it was a sticky point. "I would have to say that it would," Novick finally ventured.

Novick launched into an explanation, which Berg promptly interrupted. "I think," he said, "that we're not going to try out all the scenarios here. I think we'll have to move on."

Berg prodded the discussion along relentlessly. By the time he called for the next vote, a portion of the group wasn't even clear what the issue was, although it definitely had something to do with prokaryotes. "All those in favor?" Berg asked.

Stanley Cohen was on his feet again. "Let me make just one very fundamental point," he began.

"Stan!" said Berg. "I think what we'll ask is that you give us *your* version and—"

"Wait," said Cohen, "what is fundamental to this is—"

"Stan," said Berg. "I think we'll have to come back to this."

The discussion moved on to viruses, which itself offered little comfort. Even though the viral group recommendations had been made considerably more rigorous, the first observation by the virus group leader was that their statement about eukaryotic DNA possibly containing viral sequences had been deleted.

It was the matter of cryptic viruses again—a reflection of just how little anyone really knew about what actually resides within the massive set of information represented by a mammalian genome. The matter was then, and would remain, thoroughly cryptic itself. "Perhaps I should explain that," Novick said. "There was a discrepancy between your recommendation and the recommendation of the eukaryotic DNA group and we felt that we should take their recommendation to divide the vertebrates. And then we modified *their* decision a little by drawing the line at cold-blooded versus warm-blooded, rather then mammal versus nonmammal, since the avian [bird] viruses are konwn to grow in and transform mammalian cells."

"It's also fair to point out," someone else in the audience added, "that there are tumor viruses in frogs."

Berg wisely cut off the discussion of various animal viruses, which in meetings to come would grow so complex as to sound vaguely like some modern-day Noah stocking his Ark. But the talk then immediately encountered, once again, the notion of proscibed experiments.

"Might I ask your committee how they propose to deal with the question of type-VI experiments?" inquired a middle-aged Englishman. "It has been put to me by people in the U.K. that at this time they can see no possible experiment, with our current state of knowledge, involving, say, smallpox DNA virus and these types of vectors. It's not that they couldn't *design* such experiments—it's just that the combination of risks and benefits leads them to put them in a type-VI category."

"But that's like doing a *head* transplant," another English researcher muttered softly to a colleague. "It's simply something one wouldn't *do*."

"I thought," said Berg, "that we'd agreed to indicate that there was a split, a difference of opinion among people at the conference as to whether certain experiments should be judged not permissible at the present time."

The English researcher continued to stand. "Could we, perhaps, for my benefit as much as anyone, test that feeling? Perhaps there *isn't* a split. Could we test?"

"There have been three groups that have avoided this issue," someone agreed. "It's really a matter of principle." An amendment to the early portion of the statement was read, emphasizing that "there are organisms that could be created with current methodology that comprise risks of such intensity that such organisms ought not to be created at the present time."

"Can I talk to this, please?" Ephraim Anderson asked. "While I disagree with a good deal of the plasmid group's pronouncements, the one thing I did *not* disagree with was that there are certain organisms with which we should not *start* experimental work of this sort." Anyone he said, could sensibly compose a list of those organisms. "They are hot, and should not be worked with."

"Okay," Berg said finally. "One view says that there are experiments that should be performed only in the highest containment facilities available today. And the second point of view says that there is a class of experiments that should not be done at all with present methods."

Berg didn't even need to count hands. "Well," he said, "it's quite clear that the majority supports the latter. If we have time, perhaps we can come back to this section." He paused for a moment, then moved on. "Please keep your comments brief."

Most of the comments, for the moment, were on the order of "Is *'low risk'* in quotations one thing and *low risk* without quotations another?" or minor amendments of wording likely not even heard by half the restive audience.

Specific cases rose regularly. "I'm very puzzled," said someone. "Does this mean that doing transformation with SV40 is low risk but that Josh Lederberg has to work with moderate risk for *Bacillus subtilis*?" And feathers were regularly ruffled. "I must say, personally, having worked for four months on the plasmid document, that it's been prostituted. And that's all I have to say." But Berg and the other committee members managed to keep the whole precarious process rolling, prompted by as much reassuring language as they could muster at that hour: "I think we'll have to explore that"; "We'll certainly take that into account"; "We'd like to have a written memorandum on that."

The paragraph on virus work passed with many abstentions but only a handful of active dissenters. The section on eukaryotic DNA, however, encountered trouble right at the outset. This was, of course, the arena of yeast cells, fruit flies, human beings, and "shotgunning" —that blend of biochemical scalpel and gene amplifier that may someday allow a deep new look into the precise way individual genes are strung together on the chromosome.

But shotgunning was then very new, and the eukaryotic DNA paragraph in the statement was brief; it was something of a calm before the storm. In the months to come, the whole matter of assessing risk during shotgun experiments would grow intensely controversial, revolving around the question of exactly how one knows whether one has cloned any hazardous genes or not in a given experiment.

At Asilomar, however, the most controversy rose over another matter altogether: "Experiments with a high risk," read one sentence, "include the fusion of eukaryotic or prokaryotic genes to prokaryotic vectors when the resultant organism is likely to express a toxic or a pharmacologically active agent."

"Insulin," Don Brown, the acerbic head of the eukaryotic DNA group, pointed out, "could be considered a 'pharmacologically active agent.'" And insulin production by bacteria had been, since the outset, one of the favored examples of a fairly immediate and practical benefit of the new work. While a high risk rating would make that application very safe, it would also make it very difficult, and so Brown suggested the deletion of the reference to pharmacology.

At that point, of course, no one even knew whether one could really genetically program a bacterium like *E. coli* to turn out insulin—much less whether that hypothetical insulin-generating bacterium would prove a threat should it make itself at home in human guts.

Berg tried to maneuver around the insulin issue by explaining that the high-risk rating was meant more generally.

"Some agency might take that very literally," Brown cautioned. "The word 'toxic' would be sufficient."

David Baltimore came back to say that the phrase specifically *meant* insulin genes. "Insulin produced endogenously in the intestine, in any reasonable amount, could well be pharmacologically active, and we'd want to know a whole lot more about that before such an *E. coli* was produced. That was our judgment."

That drew the line of argument sharply. Berg—who was by then getting fairly deft with compromise statements—came up with another: "Why don't we just say 'likely to express an agent that is toxic to the organism'?"

Everyone was briefly happy until it became clear that that could still include pharmacologic substances like insulin or growth hormones. Before Brown could respond, the issue was further muddied when someone else pointed out that there was no distinction in the statement between genes that would express, and those that weren't intended to. And since no one had gotten a eukaryotic gene to express in a prokaryote in the first place, the latter was what most people were doing anyway.

"By putting these experiments," James Watson said, "which are difficult enough in the first place, in the high-risk category—which means facilities like Fort Detrick, which is probably filled up with something that is a total waste of time—we seem to be discouraging them altogether." Watson then asked an English researcher, who was already active in the field of pharmaceutical applications, what he thought. The researcher seemed to think that he, and most laboratories, wouldn't be too put out doing the work under moderate containment.

"But not high?" Watson asked immediately.

"Not high," the fellow said.

Berg looked uncomfortable; things were getting out of hand again. "I think we're going to have to move on," he said. "We're coming up on twenty to twelve."

He recognized a young American. "I would like to raise this issue, about where the line is drawn between what is low and what is high. It seems to me that there are invertebrates—take mosquitoes, for example—that contain diseases that are dangerous to human beings, and if people go willy-nilly joining mosquito DNA with *E. coli*, I'd hate to think I could get malaria from walking around on the street."

There was a quiet disturbance in the front row. "Huh?" "How could you get malaria . . . ?" How could you . . . ?"

The audience was silent. "There's no response," Berg said. "You're right, taking field-caught mosquitoes and extracting DNA and trying to clone it would be irresponsible."

"I don't know where to draw the line," the young American said. "I just ask to consider this . . ."

Berg tried to move the discussion forward before the disturbance in the front row exploded. "Can they explain . . . ?" "*Say it!*"

"Can you explain, please," a young Englishman inquired loudly, "just how a protozoan parasite could result from an *E. coli* and thus cause malaria?"

"We will take the comment under advisement," Berg said quietly.

The English researcher sat down, shaking his head. "It's bloody nonsense."

While the comment had drawn on an unfortunate example, it nonetheless underlined the spectacular number of hypothetical situations one could conjure up for recombinant DNA. The marvelous universality of DNA, that had allowed the experimenters to move from sweet peas to fruit flies to phage, had now come home to roost. At this level, the implications of DNA manipulation were really just about as broad as all of nature.

Someone immediately inquired about plant viruses. Someone else raised a more general inquiry. "I'm afraid we're going to have to vote on this," he said, "but I'm uncomfortable because I don't know what we're voting on. What do we *mean* by this phrase 'rigorously purified DNA fragments'?"

Berg showed a trace of annoyance. "We're talking about a matter of scale, not definition. I think 'rigorously purified' is different from the whole bloody genome—*you* know that, *I* know that, *everybody* understands that. We can't sit here and define it. We have to take it that the final decision on any experimenting will depend on the experiment and the peer review involved."

It would not be the last time Berg would hear about "rigorously purified," but his response was good enough to keep things moving. "I therefore call," he said shortly, "for a show of hands of those who support the version as amended and commented upon."

An elderly foreign researcher interrupted and, in a heavy accent, made a comment that soon had the English in the front row chuckling with delight again: "That's like putting mosquito into *E. coli* makes a protozoan!"

Berg ignored the interruption and asked, "All those in favor?

"All those opposed?"

There was little opposition, again, but many abstentions. "It's clear," said Berg, "that the sentiment is overwhelmingly in favor."

Joshua Lederberg stood immediately. "I don't think that's clear at all, Paul."

"Well," said Berg, looking at once harried and innocent, "there were two or three hands, and as far as I could tell, that would be overwhelming. I think we'll have to go on to item four."

Lederberg remained standing. "May I have ten seconds, Paul?"

"Ten seconds."

"Under the pressure of time and circumstances that I can deeply sympathize with you about, very complex issues are being railroaded through—"

"Nope," said Berg.

"—and the assumption is being made that there is consensus on issues that have simply not been ventilated. If you're willing to say, in your preliminary document, that that characterizes the nature of the consensus here, then I could go along. Period."

"Okay," Berg said briskly. "Then we'll move along to the next section." Then he relented, and addressed Lederberg directly. "I welcome and invite you to write us a statement as to your reservations. I think we would like to consider them in formulating the final report. But I have to take a show of hands as an expression of the sentiment as we've discussed it over a period of three and one-half days. I saw two or three noes, and I didn't count the number of ayes. So I have to call that overwhelming."

But Lederberg had clearly opened up new territory. "Could you just ask," one researcher immediately requested, "how many people abstained at including high-risk rating for insulin?"

Berg looked pained. "How many people did not vote?"

"A small number." But not, however, that small. Berg stared briefly at the audience. "Let's have the show of nands again. *Please!* There's no point in abstaining. What are your sentiments? All those in favor?"

By then, however, no one was quite certain what the vote was about. The discussion flailed briefly and then Berg asked Don Brown to reread his amended version of the eukaryotic DNA section.

Brown stood. "The first sentence is left intact—"

An immediate chorus of noes arose. It quickly developed that virtually everyone had a different idea as to precisely what the amendments had been. Finally one of the organizing committee members suggested that maybe the whole effort was hopeless, since there were amendments that weren't written down on *anybody's* sheets.

It looked for just a moment as if the whole effort might tip over into sprawling chaos. Half a dozen people started to speak simultaneously. Lederberg fumed silently; Don Brown, standing, looked incredulous; Berg looked exhausted; and then suddenly Sydney Brenner quietly interrupted.

"Could I ask whether," he inquired slowly, "we might jump to section 6? In the paragraph that states, 'This document represents our best assessment of the potential biohazards . . . ,' I would like to change the word 'best' to 'first.' "

There was a brief silence, and then, for the first time in two hours, sustained laughter.

"I think," Brenner continued finally, "that we should emphasize that these are extremely complex issues and that we may be wrong. I am very sympathetic to the minority points of view and I am not prepared to say that this is absolutely the right way to do it. Perhaps that would satisfy Professor Lederberg—that at least one member of this committee is demonstrably in favor of the view that there are, in fact, uncertainties."

In the silence that followed Brenner's bit of diplomacy, Berg—like a quarterback seeing daylight—moved to the next section of the document.

The remainder of the provisional statement was, in contrast, almost totally uncontroversial. It dealt with implementation—education, review boards, and the like

—and once the discussion moved off the territory of specific work, the level of controversy dropped and the organizing committee once more found themselves the major contributors.

In the midst of section 4, the noon lunch bell rang. Berg moved immediately to the last section—a brief paragraph that expressed some of the gaping questions that remained in the initial risk assessment, from the basic question about eukaryotic programming in prokaryotic organisms, to the provocative notion that perhaps free DNA molecules themselves, outside of their biological containers, could infect plant or animal cells.

Nobody was about to suggest any more questions. And thus Asilomar commenced a curiously anticlimactic winding-down. The eldest of the Russian delegates, whose principal contribution up to that point had been the clicking of a tiny flash camera to photograph slides during the technical sessions, rose to bless the conference in rambling fashion, concluding with a slightly garbled metaphor about scientific recombination between his country and the United States.

Very quickly, the second lunch bell rang in the courtyard outside the chapel. "If I'm correct, then," Berg said, "we've proceeded through the whole document, by the second lunch bell if not the first. And so I would like, in a sense, to formalize our actions here, by asking for a show of hands of those people who support the entire document, including the amendments we've discussed individually. All those in favor of this as a provisional statement, please raise your hands."

"Paul, I can't vote on that," said Stanley Cohen plaintively, "until I see the wording of what we're supporting."

Berg ignored the interruption. "All those," he said again, "please raise your hands.

"All those opposed to that statement."

Somewhere around four hands rose in opposition. "Okay," said Berg. "Again I would say that there is substantial agreement here on that point."

Daniel Singer stood then to "express a view of these proceedings from the outside." Regardless of the ultimate outcome, he said, he was uplifted by the way the community of scientists had organized for social responsibility. "I think you have embarked on a continuing obligation," he concluded, "with respect to this particular problem of recombinant DNA molecules, and you have set a very high standard indeed. You are a tough act to follow for other groups of scientists who must necessarily embark on this same kind of exercise in the future. It's been nice being here."

Daniel Singer had expressed it well: the Asilomar conference would indeed be a hard act to follow. And the act, as it developed, had only just begun.

8
Risk/Benefit

On February 28, just one day after the hectic conclusion of the Asilomar conference, Paul Berg, Maxine Singer, David Baltimore, and Sydney Brenner faced a bank of reporters and television cameras in the San Francisco Press Club and announced that the conference had produced what, under the circumstances, could only be considered a happy ending.

The spirit of that press conference was blithe indeed; few of the local reporters, at that point, had any idea what was involved, and the organizing committee had by then become a walking study in long-term sleep deprivation.

When Brenner described the notion of self-destructing biologic vectors, one middle-aged reporter demanded to know what possible benefit there might be for mankind in bacteria that can live only in the laboratory.

"My goodness," Brenner said mildly, shaking his head and smiling bemusedly, "I'm afraid that you really don't understand at all."

The reporter pressed his point: These microbes supposedly *explode*?

Brenner's smile never faded. "Well," he said carefully, "they don't generally make a loud noise."

In the end, however, the organizing committee answered even the reporters' most opaque inquiries with the same patience they had shown at Asilomar. But there was still a certain heady, school's-out feeling to the press conference—a feeling that seemed to carry over to the first meeting of the brand new National Institutes of Health committee now delegated to draft specific guidelines from the brief Asilomar statement.

That meeting started even as the press conference wound up, just a few blocks down the street from the Press Club in an ornate hotel conference room. There, around a polished walnut table, sat fifteen researchers and administrators—the NIH Recombinant DNA Molecules Advisory Committee. Only half of those original members remained when, more than a year later, early in the summer of 1976, NIH finally released the guidelines, setting off a storm of controversy.

But the first meeting was fairly placid. The questions at hand seemed predominantly administrative—funding standards, the establishment of local committees, methods for publicizing safer host/vector systems. The chairman of the committee—a white-haired and dignified NIH Deputy Director named DeWitt Stetten, Jr.—was so competent and charming a diplomat that most of the potential difficulties appeared altogether tractable in that good-natured San Francisco atmosphere.

There were a few hints, however, of the problems to come. Early that afternoon, for example, Jane Setlow, from Brookhaven National Laboratories—then the only woman on the NIH committee—in the midst of examining the freshly copied Asilomar statement, started to laugh.

Some of the other committee members stared at her across the polished table. "What is it?" Stetten wondered, after a moment.

"Well," she said. "I just realized that, according to this, human sex has become a high-risk experiment."

There was brief silence, then laughter. Maybe, someone suggested, it should specify "genetic recombination in the laboratory."

"But that still rules out sex," someone else said.

Setlow glanced up. "Only," she said, "in the laboratory."

Revisions ensued immediately, and in the final guidelines it required more than forty words simply to define "recombinant DNA." But already implicit in that early exchange was the broad territory that the regulation of the new technology would have to cover.

It was unknown turf that had already captured a small portion of public attention—a portion that would grow rapidly larger and more vocal in the months to come. It had captured my attention, as well, and it soon became clear that Asilomar was not an isolated news event. It was, instead, the beginning of an unprecedented process of scientific self-regulation that dealt, in turn, with a technology of potentially unlimited power.

In the months following Asilomar I started to sit in on the meetings of the new NIH Recombinant DNA Molecules Advisory Committee and also to ask more questions, in various labs, about precisely what was at stake in the new technology. The tentative and uncertain business of searching for proper controls and drafting guidelines seemed larger than the science—far more complex and critical decisions will doubtless be demanded long after the first restriction enzymes are retired to some biological hall of fame—and yet the science was unquestionably integral to the process.

Where those two components—research and regulation—meet most intimately, and often most mysteriously, is in the statistician's delight called risk/benefit analysis. Risk/benefit analysis was recognized early on to be the name of the recombinant DNA game, sprung naturally from a previous yet still controversial new technology: nuclear power. In the months following Asilomar, critics of recombinant DNA work would regularly invoke unflattering images of the nuclear industry. And the similarities between the two technologies go back even further, to the earliest days of wartime nuclear physics. One parallel is political: The only modern voluntary moratorium in the field of science prior to publication of the Berg group's letter in 1974 occurred in April 1940—before the onset of hostilities—when Allied researchers meeting in Berkeley, California, agreed to an international ban on the publication of nuclear data.

That early decision was, of course, understandable. The second parallel is slightly more eerie, however. In the midst of the American Manhattan Project, discussion arose among researchers regarding the odds on "atmospheric ignition"—the nightmarish notion that the intense heat of the first nuclear fission device might actually set the earth's atmosphere on fire, destroying oxygen in a chain reaction that would, within minutes, suffocate the planet.

Alfred Compton, a principal in the bomb project, told Robert Oppenheimer that if the theory was correct, the work should be stopped altogether. "Better to be a slave under the Nazi heel," he said, "than to draw down the final curtain on humanity." One major participant in the project quoted odds of three in a million, and an unofficial betting pool provided considerably more variation on those numbers. As late as 1945 the notion remained sufficiently unsettling that senior staff members —right up to the first detonation at Alamogordo, New Mexico—did not mention it to younger researchers.

The atmosphere, of course, did not ignite. Yet it required that first detonation to confirm the fact. As Robert Oppenhheimer observed regularly during the Manhattan Project: "There are no experts. The field is too new."

Thirty years after Alamogordo, Oppenheimer's quote accurately summarized the state of recombinant DNA engineering. But unlike the Manhattan Project, which was at once propelled and rendered sacrosanct by the war effort, the work on recombinant DNA was, almost instantly, stalled and vulnerable.

In the year following the conference at Asilomar, a handful of molecular geneticists found themselves asked to publicly justify every facet of a technology that they had restricted themselves from investigating in the first place. It was the sort of catch-22 that the nuclear power industry had neatly sidestepped—by the time the public had recovered from post-Hiroshima guilt and its associated "swords-into-plowshares" publicity and started to ask questions about the long-term wisdom of nuclear power, the principals involved had at least developed some solid statistics from which to extrapolate.

Not so, of course, the molecular geneticists. And so how does one defend a technology that has not even really started? How does one balance risks against benefits when one can't firmly describe either category in the first place?

Inevitably, one makes assumptions—about both risks and benefits. But once one begins to make assumptions in science, the field of contention is suddenly wide open. And, all too often, every assumption cuts two ways.

For example, in molecular genetics, "expression" refers to the process whereby the information coded within a gene actually produces its specific product within the cell. And early on in recombinant DNA work, it grew clear that a critical question would be whether genes plucked out of some mammal's cells could still express when transplanted into a simple bacterium like *E. coli*. During the long months of deliberation over the guidelines, that question remained essentially unanswered. While most researchers had opinions, evidence did not arrive until early in the summer of 1976, when the first published report indicated that a DNA sequence from yeast, a eukaryote, had functioned within *E. coli*. And even that left the possibility of expressing DNA from mammals only slightly less tentative.

Well before that first successful yeast–*E. coli* experiment, however, the transcription and expression of eukaryotic DNA by prokaryotes was already a major issue on both sides of the public controversy. Late in 1975, for example, Joshua Lederberg published an article in the American Medical Association's *Prism* that described the potential use of recombinant DNA techniques to produce animal proteins in simple bacteria. Lederberg's suggestions were in fact strikingly original, going past the commonplace example of insulin to include such materials as antibody globulins that might be administered to aid the body's own defense against diseases ranging from influenza to rabies. "The most important products," Lederberg wrote, "are those that remain to be discovered."

Shortly after that article, the group of young, radical scientists called Science for the People, centered in Cambridge, Massachusetts, published their own scathing attack on the wisdom of recombinant DNA work. Central among a number of unpleasant scenarios they envisioned was the possibility that bacteria programmed with insulin genes for the pharmaceutical industry might colonize some unsuspecting human gut with results, according to the article, "that would swiftly be fatal."

What seemed most striking about both Lederberg's elegant panegyric and Science for the People's ominous projection was that each was based on the same unproven assumption about the accurate expression of eukaryotic DNA in a prokaryotic host. They represented, in short, opposite opinions about a phenomenon that did not necessarily exist.

Ironically, Lederberg had been one of the researchers at Asilomar most reluctant to saddle the new technology with promises of specific benefits. The notion had made eminent sense at the time, but a few months later it became clear that when the public asks why one wants to do something already indelibly marked "hazardous," it's simply not enough to reply that it's important.

Not enough at all. The issue of recombinant DNA, which had started so quietly with the low-key media coverage of Asilomar, was about to run full tilt into the maelstrom of unbridled public opinion. In the months to come it would become a hot topic, argued about all over the country, from the air-conditioned sanctity of an NIH conference room in Bethesda, Maryland, to a pair of booths in the summer heat of a Cambridge, Massachusetts, street fair. Everyone from Science for the People to Friends of the Earth and the Democratic Committee of Washtenaw County, Michigan, would have an opinion, and the waters—unclear at the outset—would grow muddy indeed.

Amateur risk/benefit analysis almost certainly never had such a field day, and so I decided to try a bit of it myself. The result was six thoroughly baffling months, during which my only consolation was observing that the NIH Recombinant DNA Molecules Advisory Committee seemed, in many ways, equally baffled. But then that's another chapter altogether; the issue at hand is risk.

9

Biohazards

The only recorded civilian biohazard fatalities—the deaths of individuals who were in no way connected to a laboratory—occurred in London, in March 1973.

The accident started on the second floor of the London School of Hygiene and Tropical Medicine, in an overcrowded and untidy room called the Pox Virus Laboratory. At some point on February 28 a young technician named Ann Algeo, who worked in the Farmers Lung Laboratory just across the hall, visited the Pox Virus room and lingered briefly to chat with a fellow worker.

The second technician was just then harvesting newly grown smallpox virus from the half-open, fertile eggs within which viruses are customarily cultured. Due to the universality of smallpox immunization, the work was done as if it was of minimal hazard—on an open laboratory bench, disregarding most standard precautions. Seven months earlier, the technician himself had punctured his finger with a hypodermic containing concentrated smallpox virus, with no adverse effects.

Ann Algeo stood behind the technician that day and watched for a moment. Neither thought anything of her brief visit, even when, two weeks later, she awakened on a Sunday morning with headache, backache, vomiting, and fever.

It appeared to be influenza, and she was excused from work the next day. Her physician prescribed rest and fluids. By Thursday she was no better, and a rash had developed on her feet. By Friday the rash was widespread—which Algeo attributed to an allergic reaction to orange juice—but her physician, fearing meningitis, had her admitted to a hospital.

The hospital physician noted that the rash might have been an antibiotic reaction and assigned her to a general ward. By evening, her rash was generalized and her fever even higher. By then, however, the attending physician was aware of her occupation, and ordered tests to investigate the possibility that she had contracted a fungal infection from the Farmers Lung Laboratory.

The tests were negative, but by the next day, Algeo had developed raised pustules on her chest and back. Her physicians were thoroughly baffled. A tentative diagnosis of "glandular fever" went onto her charts.

That same day Ann Algeo started a rapid convalescence and soon felt sufficiently sociable to swap conversation and reading matter with an elderlly Irish widow in the bed next to her. The widow lent her a copy of a newspaper called *Ireland's Own*. Later that day, Algeo passed it back, and the widow passed the paper on to two of her visitors—her thirty-four-year-old son and his wife.

Ann Algeo was released soon after; her last remaining symptom was only a small oozing pustule on her right hand.

Not until a full week later was it recognized that Ann Algeo had classic smallpox. She had apparently carried the disease out of the laboratory and had been infectious —"shedding virus"—since the middle of March. A massive public health operation went into gear immediately, but it was not until April 4 that the Irish widow's visitors were identified as possible contacts. And by then, her son and daughter-in-law had already spent two days in another hospital—with a tentative diagnosis of food poisoning.

Food poisoning was, of course, no more the young couple's problem than influenza had been Ann Algeo's. But the virus treated the married couple, whose immunizations were probably less recent than Algeo's, with far less mercy. The wife, extremely ill on admission, died quickly from an uncommon form of smallpox that causes such severe bleeding, both internally and from the skin, that it can be misdiagnosed as a blood ailment. Her husband died a week later, with high fever and body lesions so generalized that, at the end, it was difficult to distinguish one outbreak from another.

One body was cremated immediately. The other was sealed in a coffin filled with disinfectant-saturated sawdust and subsequently buried by freshly vaccinated gravediggers. The area of the cemetery itself was not used again for six months.

The renegade pox virus was apparently buried there also, and for that, the public health officials involved in tracing the outbreak breathed a collective sigh of relief. As the official report on the incident suggested, had the outbreak not been contained with the luckless young couple, "this could have led to a large number of third-generation cases."

The London accident involved a disease that has been known to man for hundreds of years, and for which effective vaccination techniques have existed since the eighteenth century. The accident occurred, ironically, just eighteen months before a target date set by the World Health Organization for the total eradication of the disease from the planet.

When a well-known human pathogen can cause such tragedy and near disaster, it seems little wonder that the notion of the accidental release of new infectious organisms containing novel genetic material raised concern on every front from the National Institutes of Health to the Cambridge City Council. After smallpox, many wondered—what next?

Yet when the question of biohazard control and recombinant DNA first attracted the attention of the national media, in June 1976, not even the *New York Times* managed to get the story straight. By then, through some bizarre media alchemy, recombinant DNA work had become "creation of life" experiments. In the frequently quoted words of Alfred E. Vellucci, mayor of Cambridge, Massachusetts, "They may come up with a disease that can't be cured—even a monster! Is this the answer to Dr. Frankenstein's dream?"

Mayor Vellucci's assessment, while perhaps not coolly reasoned, was significant nonetheless. It seemed that Harvard University, after considerable interdepartmental battle, had finally approved plans for the construction of a biohazard containment facility—a form of laboratory designed to contain potentially dangerous microorganisms, and one that would be used in part for work with recombinant DNA molecules, as required by the NIH guidelines. A local newsweekly, however, had picked up on the plans with a front-page story headlined, in bright orange type, "BIOHAZARDS AT HARVARD."

"Scientists," read the subhead, "are on the brink of undertaking revolutionary genetic research which creates new life forms—and dangers which the public knows little about."

The latter problem was soon to be remedied. While the technical portions of the lengthy article were not particularly clear, the writers' concerns most certainly were. The research sounded murky and risky, and the Harvard researchers who supported the new facility came out as either hollow-headed morons or experimental zealots willing to wipe out most of eastern Massachusetts in the quest for knowledge.

Mayor Vellucci took immediate notice, and his City Council subsequently voted 9–0 to hold a public hearing on whether Harvard should be allowed its facility at all. The hearing was scheduled, ironically, for the same day that the final NIH guidelines were to be issued, 700 miles south in Bethesda, Maryland. The recombinant DNA controversy, with special emphasis on biohazards, was about to go public.

That same week I was on the other side of the country, sitting in a small auditorium at the University of California at Los Angeles Medical Center, on the fourth floor of a massive building whose bulk is surpassed only by the Pentagon.

The occasion was a three-day NIH short course called "Biohazard and Injury Control in the Biomedical Laboratory." It was a small gathering that hot southern California week—about thirty people in all, ranging from young women just out of their teens to gray-haired senior research directors, from Ph.D.s who run million-dollar research programs to the assistants who wash their lab equipment. Their home laboratories ranged from Phoenix and Seattle to San Diego and Berkeley.

All of the students, however, had one thing in common: They worked at the most hazardous fringe of biological research, with infectious agents either suspect, dangerous, or lethal. It is an experimental arena that has, since the turn of the century, yielded more than 3500 cases of disease and over 150 deaths among laboratory workers. The new directions of modern biology have lately created even newer hazards, more ubiquitous and more insidious. And thus the point of the three-day course was simple: to keep its students inside the laboratory but out of the statistics—and, past that, out of trouble in the community.

The first part of that charge involved a good deal of self-interest. "Many of the diseases which we now know to be infectious," a biohazard expert had told me earlier, "were determined to be so only because of laboratory accidents. In a sense," he said cheerfully, "the laboratorian is a sentinel for hazards that may come in the future."

No one at the UCLA class had any intention of serving as a sentinel. The literature of laboratory-acquired infection already teems with more than enough bizarre diseases—a nightmare assortment of illnesses caused by, in the cooler language of the laboratory, "agents pathogenic to man."

Take, for example, one discovery by a "laboratory sentinel": Marburg virus—a previously unknown disease that welled up out of a batch of laboratory-bound African green monkeys during the summer of 1967 to afflict thirty-one humans and kill seven in Marburg, Germany. The virus killed its host monkeys, as well, and even now the source of the disease remains a mystery. Whatever its source, however, the virus proved spectacularly contagious for humans—a "micro-epidemic"—infecting everyone from the monkey handlers to their doctors and nurses, a morgue attendant, and one survivor's wife, who contracted the virus herself almost three months later via her husband's semen.

Marburg virus disappeared as mysteriously as it arrived—recurring only once, nearly a decade later, in eastern Africa. Not so, however, Q-fever—a curious, highly infective illness of Mediterranean origin that has caused nearly six hundred laboratory infections in recent decades, including twenty-three fatalities. Q-fever, a respiratory ailment that can also cause liver and heart damage, is caused by a class of bacteria called rickettsiae, microorganisms so hardy and infectious that by current estimates, even a single one of the microscopic cells can cause Q-fever in a human. By contrast, it generally requires fully a million *Salmonella* bacteria to induce disease. And Q-fever travels well: When a shipment of laboratory guinea pig carcasses, infected with Q-fever, was sent to a rendering plant, thirty-five workers promptly contracted the peculiar fever.

The nastiest biohazard thus far, however, seems to be something called Simian Virus Type B. Type B was the first recognized monkey virus (although it doesn't seem to particularly bother its simian hosts) and also the first monkey virus found to infect man.

And infect it does. "No clinical evidence of disease" is the way most B-virus infected monkeys are described in the literature. Most humans infected with the virus are diagnosed as having "acute ascending myelitis"—an incurable, progressive central nervous system degeneration that has, by now, claimed the lives of seventeen of the twenty suspected cases.

Some of those infections were insidious indeed. In one case, the saliva of a rhesus monkey came in contact with a small wound on the palm of a worker. In another, a technician cut his hand on the broken glass of a laboratory flask that had earlier held monkey kidney cells. A third case, still unexplained, involved a student who was simply preparing a monkey skull for study.

Marburg, Q-fever, and Type B represent only a sample of laboratory-acquired illness; the full list reads like a Cook's tour of human misery, from familiar old scourges like plague, cholera, syphilis, anthrax, polio, and yellow fever to exotic new infections like Yaba virus, West Nile fever, Venezuelan equine encephalitis, and Lassa fever. Yet curiously enough, none of the thirty-odd researchers attending the Los Angeles biohazards class had anything to do with the above agents. Their concern, for the most part, was a newer and vastly more subtle hazard: the oncogenic viruses.

The great American search for a human cancer virus has been under way since 1966, and while it has thus far produced no lasting candidates, it has unquestionably introduced some unsavory viral characters into the laboratory menagerie.

Simian Virus 40 is one of these. SV40, of course, was the monkey tumor virus accidentally administered to several million young Americans along with their polio vaccinations in the early 1950s, and while that's an extensive field test, SV40 is still considered a dubious commodity. More dubious still are some of the other animal tumor viruses—feline leukemia virus, say, or woolly monkey fibrosarcoma—which have already displayed the distressing ability to cross "species barriers" and cause disease in animals far removed from the original host. The C-type RNA viruses are capable of integrating their spurious genetic information into the host genome so seamlessly that a whole virus can persist harmlessly within the creature, be passed on from generation to generation, and then emerge years later, unannounced, once again whole and infectious.

If there is anything at all thus far established in recent virology, it is that viruses can't be trusted. An excellent example is the relatively new class of "slow" viruses—

the cause of diseases like kuru, a fatal central nervous system disease that was apparently transmitted by the ritual ingestion of human brain tissue by New Guinea cannibals. A victim of kuru could live from many months to as long as several decades before symptoms arose. A related disease in sheep sometimes infects animals who will die of old age before the symptoms of the illness even have a chance to appear.

"Only old people," one speaker jocularly advised the biohazards class, "should work with the slow viruses." The observation drew few chuckles from the assembled researchers. They were after all, working with viruses of uncertain hazard whose ultimate effects, like that of the slow viruses, might not appear for years.

Can one "catch" cancer in the laboratory? There has been one apparent death from laboratory-acquired cancer—a French medical student who, in 1923, punctured his hand with a syringe filled with fluid from the site of radical mastectomy. Two years later, a large tumor developed at the injection site, and a year after that the young man died of widespread malignancies.

Now, more than fifty years later, that is still the only recorded case. Especially in recent years, there have been suggestive anecdotes—a young cancer researcher, for example, who brought home some laboratory mice for his baby son to play with, and a few months later the child died of a brain tumor—but isolated anecdotes are not data. One of the major cancer virus research centers in this country requires its workers, before they enter the laboratory, to don pea-green coveralls displaying a colorful embroidered patch—a round diagram representing potential causes of cancer. Virus particles are depicted at twelve o'clock, a human figure at three o'clock, a beakerful of chemicals at six o'clock, and a mouse at nine o'clock. Precisely in the center is a radiation symbol. All of the symbols save one are connected to each other by solid lines: chemicals can cause cancer in both man and mouse, radiation does the same, and virus can cause cancer in mice. Only one dotted line, plus an embroidered question mark, connects the virus particles and the human. When the patches were ordered, some years ago, no one knew. And no one knows today.

But no one really wants to take chances. During a question-and-answer session at the UCLA class, a technician asked about a type of disinfectant gas that her lab used to decontaminate areas that had contained oncogenic viruses. When she mentioned the name of the gas, a low murmur went through the class: the gas had just recently been identified as a potential chemical carcinogen, or cancer-causing agent.

The speaker passed on the new information. "Seems kind of ironic," he observed, "to use a carcinogenic disinfectant to protect yourself against an oncogenic virus."

No one laughed. Someone in the class promptly pointed out that there really wasn't much data available on just what works to inactivate oncogenic viruses. There was a brief silence, and two of the class organizers—both experts in biohazard control—glanced at each other in the front of the lecture hall. "Well," said one finally, reluctantly, "I guess that's the sort of study *we* ought to do. Tell you what," he said to his colleague. "I'll be in charge of the disinfectants. *You* bring the viruses."

The uncertainties facing cancer virologists today are probably an accurate foreshadowing of the future of biohazard control in the first tentative years of recombinant DNA work, when no one can be quite certain what a given experiment might produce in terms of human pathogenicity. During World War II biological warfare research, on the other hand, laboratory workers tended to be very careful; whatever organism one's culture dish contained, it was there solely because it made people sick. And thus it's not surprising that much of the technology of biohazard control originated at Fort Detrick, the American center for biological warfare research that operated for nearly three decades in the green countryside of suburban Maryland.

The Detrick work was shut down in 1969 (although some of the facilities are still used for cancer virus work), and the research itself has been forbidden since 1975, the year the United States signed the Biological Warfare Convention. The defunct project is now often delicately referred to as "our biological defense program"; in fact, a strictly defensive program is still permitted under the terms of the agreement and continues at the Dugway Proving Grounds in Utah. The euphemism for the earlier research is, nonetheless, rather apt: Since most notions of biological warfare have thus far proven tactically infeasible, Fort Detrick's real contribution seems to be its efforts to protect its own workers against the relentlessly pathogenic microbes they handled daily.

Over the course of its operation, Fort Detrick recorded 423 hospitalizations and three fatalities, which have been attributed to every disease from Q-fever to Venezuelan equine encephalitis. It was standard policy for workers recuperating from a laboratory-acquired infection to receive full pay rather than sick leave.

Fort Detrick probably represents just about the worst of all worst-case situations for biohazard containment: workers handling, for nearly thirty years, the most pathogenic of human diseases, often in large quantities and often in situations dubbed "aerosol challenge." Aerosol challenge means spraying high concentrations of bacteria or virus in the direction of one or more experimental animals, and in terms of human health hazards it likely still qualifies as the single most outrageous laboratory technique yet devised. In 1959, in a single building on the grounds of Fort Detrick, 46,412 mice, 3,013 guinea pigs, 25 rabbits, 276 monkeys, and two chimpanzees underwent aerosol challenge with everything from anthrax to plague.

Thus it is really to the credit of Fort Detrick's biological safety officers that over the course of the project progressively lower levels of biohazard accidents were recorded. During the final decade of operation, no fatalities and only 52 infections were reported, many not even requiring hospitalization. The previous decade had seen an infection rate four times as high.

In part, that improvement was due to the introduction

of new equipment—most of all the biological safety cabinet. The biological safety cabinet—by now standard issue in this field—is essentially a ceiling-high box wherein all experiments are performed in the presence of a constant, filtered upward airstream that conducts any contaminants away from the experimenter. In practice, it looks like a laboratory bench that has been sealed on three sides and whose front is a thick glass sheet that permits only the introduction of the researcher's forearms.

The biological safety cabinet and associated techniques were doubtless important in the improvement of biohazard control at Fort Detrick. Important also, however, was the development of effective vaccines against some of the more popular experimental subjects—anthrax, tularemia, Venezuelan equine encephalitis—which were then routinely administered to everyone associated with the laboratory.

Vaccination, unfortunately, is likely to be of little help in the biological safety programs of the future, where the most hazardous agent may well be the one altogether unforeseen. But the remainder of the Fort Detrick legacy is now main-stream biohazard control, whose long-term dependability was one of the issues at the heart of the recombinant DNA controversy.

A few days before the Los Angeles biohazards class, I visited a small building in Bethesda, Maryland, that represents one of the purer distillations of the lessons of Fort Detrick.

Building 41, on the sprawling grounds of the National Institutes of Health campus, is not easy to find, and while most of the people who work on the NIH grounds have heard of the facility, few know exactly where it is. Moreover, once one has found the building—tucked away at one edge of the property, a one-and-a-half story structure of poured concrete interrupted only by long dark strips of tinted glass—it's another matter altogether to find the front door. Most of the building's doors have no outside handles at all, and those that do are locked and bear bright-red warnings at waist level: KEEP OUT—RESTRICTED AREA. Constructed in 1968, at a cost of $3.5 million, Building 41 was designed specifically as the first high-security containment facility for a human cancer virus.

The prime candidate then was leukemia; by now, the field of suspects has grown diffuse, and no one is willing to predict the next turn. But recombinant DNA work—particularly its application in the dissection of the tumor virus genome—suggests strongly that if the techniques go forward, Building 41 might soon find itself a busy containment facility indeed.

The man in charge of Building 41 is Emmet Barkeley, who is probably the chief biological safety officer in the country and already a familiar face at the NIH Recombinant DNA Molecules Advisory Committee meetings. Young and personable, Barkeley is a public health specialist who went back to graduate school and studied microbiology in order to learn just what his charges in Building 41 might be up to. His professional manner seemed that of the quintessential safety officer: At the committee meetings, he was incapable of answering any question briefly; his replies were always meticulous, complete, and presented in deliberate, step-by-step fashion —the same way, clearly, as he would have laboratory procedures conducted.

Barkeley had become increasingly involved with the recombinant DNA controversy. When I visited him in his office in the "no-risk" zone of Building 41, he was working on a portion of an environmental impact statement for recombinant DNA work—a document which NIH suspected, correctly, that critics would soon request. Barkeley seemed as puzzled as some of the older hands on the NIH committee: "How," he wondered, "do you write an environmental impact statement about totally hypothetical situations?"

In the beginning, Barkeley's presence at the committee meetings had been to advise on the fundamentals of physical containment, i.e., precisely what level of laboratory caution should be required for each kind of recombinant DNA experiment. It was more than a small problem, since at the outset, there wasn't even a universal rating system for the containment of naturally occurring pathogens.

The Center for Disease Control, in Atlanta, Georgia, and the National Cancer Institute both issued booklets assessing the hazards inherent in various bacterial and viral agents. CDC defined five classes of disease caused by these agents, beginning with harmless bacteria, progressing through somewhat more serious nuisances like *Salmonella,* and ending up with Marburg virus, wild-type yellow fever, and certain exotic animal diseases that are strictly monitored to prevent their entry into the United States. NCI did the same for oncogenic viruses, describing three classes: SV40 is low-risk, a small number of other animal tumor viruses are moderate-risk, and the high-risk category is vacant—reserved, of course, for the first true human cancer virus to come along. Each class of agent requires its own level of physical containment, which ranges from little more than hanging a warning sign on the lab door to elaborate precautions costing tens of thousands of dollars and requiring everything from negative-pressure laboratories to totally sealed glove boxes.

The first draft of the NIH guidelines, written during the months following Asilomar, created a new rating system for physical containment that has rapidly become a standard. Four levels were described, dubbed P1 through P4, along with three ratings for the relative safety of disabled host organisms, called EK1, EK2 and EK3. The containment requirements for all proposed recombinant DNA experiments would thereafter be described in terms of both physical and biological techniques.

In terms of physical containment, an experiment rated at P4 would be considered potentially hazardous indeed. Only a few P4 facilities exist in this country, including Fort Detrick and two laboratories built specifically to receive returning space vehicles. Building 41 is, of course, another.

In the summer of 1976, however, Building 41 was only running at P3, since the work involved tumor viruses grown in culture dishes rather than live animals. Barkeley commented that he wouldn't allow such moderate-risk experiments to be performed under the more stringent P4 conditions. "It's inappropriate," he said, "to use excessive controls when they're not warranted, the fear being that when they *are* needed the means for abusing them will already be established. You can't," he said, "cry wolf."

To enter Building 41's work area, one is required to don a green jumpsuit and cover one's shoes with plastic booties. P4 would have required a series of mandatory showers and the use of "transitional garments." The precautions, especially at the P3 level, are in everyone's interest: Animal tissue cultures, grown in glass, are notoriously sensitive to contaminants, and an accidental infection—fungal spores, say, carried in from outside— could destroy a delicate experiment overnight.

At one time, laboratory design was rather a question of walling in or walling out. When Paul Berg's virus lab was built at Stanford a few years ago, there had been some question as to whether the room should be negative pressure, so that air would only come into the lab and the virus inside would stay inside, or positive pressure, so that airborne contaminants would be pushed out.

Berg chose negative pressure and by now, with the advent of recombinant DNA, that arrangement is required for all P3 facilities. Building 41's P4 ability, however, goes negative pressure one better—all laboratories are accessible only by double-doored air locks, wherein one door locks automatically the moment the other is opened.

Past building 41's air locks lies a system of pea-green "support" corridors surrounding the actual laboratory space. Regularly spaced along these wide, fluorescent-lit hallways are small, waist-high pass-through boxes that open into the lab space itself. Each stainless-steel box has wire-reinforced glass doors at both ends and constant ultraviolet irradiation within; under the highest containment conditions, this is how laboratory instruments would be delivered—through one door at a time.

The small rectangular core of Building 41 is the actual laboratory space. At first glance, it appears not much different from other biological facilities. The corridors are lined with carts bearing clean glassware, and posters on the walls advertise everything from a local sandwich shop menu to cancer symposia and mail-order nucleic acids.

But details soon emerge: Emergency hand-washers and body showers are regularly spaced along the corridor walls. A young man collects trash with elbow-length rubber gloves, sealing each sack before he lifts it from the can and then dropping it into another sealed container. No one smokes, drinks, or eats in the work area. Each individual laboratory room has warning signs and a door equipped with a small pass-through panel for specimens.

Much of what passes into this laboratory space, of course, comes out again, usually in the form of dirty glassware or soiled laundry. There is a special exit for such material—for anything, actually, that is not supposed to be alive. This is the pass-through autoclave, a variant of the bench top autoclave, which is a common piece of equipment in microbiology used to sterilize small quantities of instruments and glassware. The bench-top autoclave is usually a thick-walled steel, container the size of a small picnic cooler, with a screw-down top. The autoclave develops an internal atmosphere of high-pressure super-heated steam—an environment somewhere around 300° which, over twenty minutes or so, effectively destroys any living organism remaining within.

Building 41 takes the notion a step farther. Here the autoclaves are a pass-through interface between the labs and the support corridors the size and shape of commercial dry-cleaning units with huge gleaming steel doors on each side. Their capacity is sufficient to sterilize anything up to the size of a chimpanzee cage—and nothing inanimate leaves the lab by any other route.

The life-eradicating ability of the massive autoclaves is checked regularly by the use of paper strips coated with spore-forming bacteria. These sporulating bacteria characteristically react to adverse conditions by creating almost indestructible outer coatings designed to keep them alive for decades and even centuries. If the sporulated bacteria die in the autoclave, the assumption is that nothing else—hidden, say, in the laboratory laundry —could survive.

Midway during our walk through Building 41, one researcher cornered Barkeley with a complaint about an excessively noisy biological safety cabinet. "It's like working in a wind tunnel," the scientist told him, "just impossible." Couldn't they, he wondered, just shut the exhaust valve down when they were doing work that didn't require such stringent containment?

Barkeley listened, nodding sympathetically. The work in question, however, involved Epstein-Barr virus—the virus thus far most suggestively linked to a form of human cancer—and so his answer was both cordial and firm: No. How could one be certain that the exhaust valve, once throttled, would be opened again? Not that anyone would do such a thing on purpose—it's just that good biohazard practice shouldn't rely on too much minute-by-minute judgment on the part of the researcher, who more likely than not would be preoccupied by the scientific details of his experiment.

The researcher looked briefly unhappy. Barkeley promised to have the safety cabinet serviced as soon as possible, and the scientist shrugged and walked back into the coffee and cigarette area.

At that point, only one room in Building 41 was idle: The narrow, semidarkened laboratory that contained the glove boxes. The boxes were really a series of massive glass tanks, set in stainless-steel frames and secured by rows of heavy hex nuts. Each was the size of a one hundred-gallon aquarium, containing one or another piece of laboratory equipment—incubator, high speed centrifuge, binocular microscope—sealed within. The only entrances were small round openings that led into thick rubber gloves. Each box would be pressurized with Freon before use to detect leaks, and one was even specifically designed to contain nothing but an atmosphere of pure carbon dioxide.

This fifteen-foot workbench cost $60,000 ten years ago, and it still represents the highest level of biohazard control. This was to be the original home of human cancer virus, but in the decade since its installation it has been used only twice, briefly. Before this decade is out, however, the Building 41 glove boxes will almost certainly be booked to capacity: Even before the guidelines had been officially released, the first P4 recombinant DNA experiment had already been proposed for Building 41.

And so the future for biohazard control—for better or worse—seems secure indeed. But I had one more question about the past: What were the origins of the biohazard symbol itself, whose arcane configuration would, in the months to follow, figure in the artwork of at least half a dozen national newspaper and magazine articles.

"That was one of our first projects," Barkeley said, "back in 1965. We couldn't use the skull-and-crossbones; it's too familiar, and not specific enough. And so Dow Chemical, on contract, developed a new symbol for us."

The symbol is effective. While the radiation symbol suggests dispersal outward, the biohazard symbol, with its interlocking partial circles, connotes pernicious infectivity.

"But I'm afraid it's already used too often," Barkeley said. "If we end up with too many signs in the laboratory —radiation, biohazard, chemical carcinogen—they may lose their effectiveness altogether." He shook his head. "What would really help with biohazard control," he said, almost wistfully, "would be if you had people working with agents that they *knew* caused disease in man."

But that's not, of course, the case with the new biohazards. And neither is there an Emmet Barkeley personally overseeing every lab in the country. Both seemed good reasons for the NCI-sponsored biohazards short course at UCLA. The course, which is given at various locations during the year, is organized by members of the

School of Public Health at the University of Minnesota —solemn, cautious midwestern academics who seemed to see their mission as something midway between education and evangelism.

"Your workbench is not your office!" one white-haired, grandfatherly gentleman thundered from the podium. "The coffeepot has no place in the laboratory!" In his view, in fact, very little seemed to have a place in the laboratory. "A ventilation designer gives you five or six air changes per hour in your lab and he thinks he's done well. In my book, he's just *begun!*"

He scoffed at P4 and suggested P5, which would involve the remote control manipulation of miniaturized experiments in a fashion similar to that used for radioactive materials. Yet he didn't respect expensive hardware by itself: The lunar receiving lab at Houston, built to handle material returned from the moon, was, he pointed out, the most sophisticated containment facility on earth, yet within twenty-four hours of its first use, there were indications that air from within the laboratory had escaped into the environment.

Good hardware, clearly, was essential. Good technique was crucial. The next speaker told the story of seven careless researchers at one of the nation's leading cancer research centers who worked with tissue samples from patients with Hodgkin's disease—the lymphatic cancer considered by some to be possibly contagious. None contracted Hodgkin's disease—but three developed herpes infections, four ran high fevers and three grew warts. Why? No one knew precisely. "But they were mouth-pipetting," the speaker noted sternly. "They don't do that anymore."

Mouth pipetting is the use of lips and lungs to draw fluids into narrow glass tubes for transfer to other vessels. It is, for obvious reasons, a classic biohazard proscription. Accidental self-inoculation is the leading cause of laboratory infection, and thus the syringe has been dubbed the most dangerous instrument in biomedicine. But the mouth, rather decisively, comes in a close second.

The biohazards class featured three folding card tables piled high with samples of laboratory safety equipment, fully a third of which were mechanical pipetting aids that don't require the application of one's lips. The reason that so many alternative devices exist is that no one likes to use them in the first place—mouth-pipetting is really a lot easier.

But then that's the sort of attitude that the short course aimed to alter. Even if it's harder, it's better, went the message: or else. And the or-elses were elaborate. For example, there was the woman at Fort Detrick who worked with herpes virus and promptly developed painful sores on the tips of the fingers of one hand. She was puzzled—she'd always worn her laboratory gloves —until someone noticed that at the end of the day, she invariably pulled her second glove off with her first, ungloved hand.

A speaker demonstrated how to remove gloves. "Glove touches glove," he said, smiling slightly. "Skin touches skin. That's not so hard to remember, is it?" He paused briefly. "And remember to take off your plastic booties *before* you take your gloves off. It's nice," he concluded, "to be wearing your gloves when a flask of malignant melanoma fractures in your hand."

Then there was the whole business of who doesn't belong in a biohazardous laboratory in the first place. The instances ranged from the obvious, such as pregnant women and children, to the more subtle: people taking steroid medications, for example, who might have reduced immunity to some random virus or bacterium. And while one would think that most people would avoid such laboratories like the plague, it's not necessarily so. After several months of operation, researchers at Building 41 found that tennis players from a neighboring court were regularly wandering through an unlocked door to use the showers at the containment facility. At Fort Detrick, a plumber who had just had a tooth pulled encountered a laboratory door that read DO NOT ENTER WITH OPEN WOUND. He entered, not realizing that his recent extraction qualified eminently as a wound, and died sometime thereafter of anthrax.

That last story provoked a brief discussion at the Los Angeles course about whether one could have one's laboratory technicians sign "informed consent" forms before starting work. The discussion ground to a halt over the question of how a researcher could inform his employees of a hazard about which he himself is essentially uninformed.

But after-the-fact responsibility was not really the point. Like all good safety procedures, there was no place here for a fallback position, and thus the emphasis for the remainder of the class was technique. A presentation called "Aerosol Dispersal in the Laboratory" featured dramatic charts of the minute droplets that radiate from a single drop of fluid falling only two feet—with each spot of scatter almost certainly containing at least one virus particle. "Assessment of Risk in the Cancer Virus Laboratory" was a slide show that began with cheery flute music behind a staid title slide and then segued into a Disney-style narrator talking, rather incongruously, about laboratory-acquired cancer. And an entire film was devoted to the proper use of biological safety cabinets: A smoke generator was set up inside the hood, and as a technician went through some standard laboratory procedures, each careless movement sent a small puff of white smoke out of the cabinet. "Remember," said a speaker after the film, "there are many factors to keep in mind while using the safety cabinet. A column of heated air, for example, rises over the average human being at forty or fifty feet per minute."

Many factors indeed. And the painstaking emphasis on technique reflected one peculiar and potentially dangerous facet of the new biohazard problem, one that would only increase with the rise of recombinant DNA work. "Ten years ago," said one of the speakers at UCLA, "we had biologists in the lab. Now we've got biochemists. And the biochemist may not always be aware of how to handle infectious agents. He'll use virus, say, in incredibly high concentrations and then do things with it that are guaranteed to create aerosols."

The region between life and chemistry is tricky territory. "Sterilization is difficult," said another speaker, "when you don't know whether the agent is living or dead in the first place. I have enough trouble with viruses—I don't even want to *talk* about the people who just use strands of DNA."

Months earlier, at Asilomar, Joshua Lederberg had expressed the same concern more tersely: "Some of these people," he'd said during a lunch break, "don't realize that what they're working with now is *alive*." After that lunch, in a nearby rest room, I'd noticed the range of hand-washing techniques. One could pick out those who had been trained to wash—probably physicians, or medical microbiologists. The rest of the Asilomar contingent, for all intents and purposes, washed their hands the way their mothers had taught them.

It was a very minor indication of a larger dilemma. Molecular genetics, with its pursuit of even simpler genetic systems to study, had managed to chase the logic of life right down to its chemical bones; and now, ironically, the very success of that reduction process made it necessary to remind a new generation of researchers that the object of their curiosity was, after all, still life.

The Los Angeles biohazards class was dismissed on schedule, and the participants were sent back to their labs, laden down with ten pounds of teaching materials and charged to continue the education process at home. Just before I left, a young woman from the Minnesota group told me that their next assignment, that fall, was to run similar courses designed specifically for recombinant DNA researchers to be held, appropriately, at Stanford and at Cold Spring Harbor. She was concerned, however, about the prospect of down-home Minnesotans bringing the gospel of laboratory safety to such luminaries of modern biology. "How do you teach a short course to would-be Nobel laureates, when you're from Minnesota?" she wondered. "Maybe we shouldn't call them short courses. Maybe we should call them workshops."

My advice was not to worry; regardless of what the

safety class was finally called, even the feistiest young lions of molecular genetics would probably listen attentively. The controversy in Cambridge seemed ample warning: The new practitioners of recombinant DNA had best rapidly develop some convincing answers to hard questions about public safety. One of the very first speakers at the biohazards class had probably made that point most clearly of all. "In the old days," he said, "failure to sterilize meant a failed, contaminated experiment. Now, failure to sterilize may mean that someone with a badge could come by and take you out of your laboratory. This is the era of community responsibility; 'getting by' is no longer enough. No longer will a white laboratory coat qualify you as a expert in biohazard control. It will only be good for selling hemorrhoid medicine on television."

10

Synthetic Muscles
and Tailored Microbes

I once read an article in *Scientific American* that contained elaborate plans for building a synthetic muscle—a small stretch of fiber that, when maintained in the proper chemical bath, would regularly contract and elongate and thus imitate the behavior of a natural muscle. The notion was instantly enchanting, and before long I'd conjured up a vision of a massive, seventy-five-pound lump of synthetic muscle pulsing regularly beneath, say, the hood of a suitably organic roadster. The idea was, of course, ridiculous; to expand the experiment described in the magazine would have required a monstrously complex set of support devices which, if not utterly impossible in the first place, would certainly be so massive and unwieldy as to immobilize permanently any kind of vehicle.

Yet the attraction of the idea is obvious: Biological systems, from biceps to brains, offer flexibility and efficiency without parallel in human technology—and impossible to ape. It becomes clear, after a bit of thought, that one can't talk about building a giant muscle; one really has to talk about *growing* one.

The whole idea belongs more to science fiction than a book about the infancy of recombinant DNA. But for me, that brief fantasy was really the first time that the implications of biological engineering struck home, a vague hint of what, in some distant future, might prove possible when the language of molecular genetics has

142

been mastered and the elaborate biochemical blueprints contained within the genes have been set to human purposes. It was, in short, the first time that I understood that the results of the biological revolution, whether we like it or not, will someday make the industrial revolution look ancient history indeed. And recombinant DNA work is the first, very tentative step toward that future.

After wading through a few texts on molecular genetics, however, it can grow deceptively easy for the layman to feel altogether chummy with the notion of manipulating the lengthy, intricate DNA molecule. The chromosome map of *E. coli*—a series of neat circles marked out to designate each gene known to control a specific biochemical function—begins to seem not much more remarkable than a sophisticated topographic map. At conferences, speakers point confidently to portions of that map and speak blithely of deleting a gene here, adding a gene there. And after a time, the business of recombinant DNA begins to seem as easy as the way one newspaper article characterized it: a matter of chemical scissors, needles, and thread.

What is easy to forget in the midst of this familiarity is that these manipulations are performed on an almost unimaginably small scale. Several million *E. coli* are at least ten times smaller than their host—far too small to be seen in even the most powerful light microscope. And the size of an individual gene sequence within any creature is really too small to imagine in the first place.

The genome is a realm where the units of weight are based on single hydrogen atoms and where distances are measured in units so minute they can also describe the length of a light wave. This is the territory of recombinant DNA and all true genetic engineering to come; simply in terms of the dimensions involved, it seems a miracle that the notion of manipulating these tiny particles is even thought possible.

One place where that notion is thought to be far more than simply possible is the Cold Spring Harbor Laboratory, a facility that many consider to be something of a landmark in the history of molecular biology.

Cold Spring Harbor is a rustic scatter of homes, cabins, and dormitories on one hundred wooded acres of Long Island. The land has been, by turns, a sheep farm, a whaling town, a resort, an enclave of exclusive estates, and most recently, the American cradle of molecular genetics.

In the winter, flocks of Canadian geese visit Cold Spring Harbor; in the summer, dogwoods flower and the geese are replaced by hordes of molecular biologists. Here, during the late 1940s and early 1950s, in summer sessions punctuated by tennis and sailing, the Phage Group first spread the notion that the long-neglected bacterial viruses might yield genetic truths applicable to creatures far larger. A quarter of a century later, Cold Spring Harbor has become a pilgrimage for molecular biologists from around the world. A few dozen who visit stay on as full-time researchers under the watchful eye of James Watson, and hundreds more attend the annual workshops that have made CSH a unique sort of academic summer camp.

Even a brief walk through the scattered laboratories of Cold Spring Harbor suggests just how broadly molecular genetics spans the line between life and chemistry. On one door hangs the biohazard sign; on the next, the radiation symbol. And the tools behind those doors range from the exotic biochemical distillates of countless million deceased bacteria to costly scintillation counters the size of home freezers.

Odd tools indeed. The cornerstone of the new technology, however, is the restriction enzymes—the bacterial defense mechanisms that exist in nature to seek out and slice up invading segments of foreign DNA. The curious properties of the restriction enzymes have been studied for more than a quarter century—but only recently have those biochemicals become scientific celebrities.

Likely the most famous restriction enzyme to date is EcoR1—the enzyme isolated from *E. coli* in San Francisco, and then found by Stanford researchers to cut DNA sequences in such a fashion that they could be functionally rejoined; the crucial discovery that abruptly made recombinant DNA a matter of public concern.

But EcoR1 is hardly the only restriction enzyme. A number of bacteria produce similar substances, and by now nearly fifty of the DNA-cleaving enzymes have been identified—each labeled with a brief, cryptic designation such as Hind III, Bam or Sal. And while not all have EcoR1's ability to leave those convenient "sticky ends," each—as in a set of wood-carving chisels—has a purpose.

In the six-foot-high incubators at Cold Spring Harbor there are fat glass jugs full of nutrient the color of apple cider, thoroughly clouded with several trillion thriving bacteria of one strain or another. Each jug is harvested at regular intervals and its contents processed to yield the specific restriction enzyme that its strain of bacteria produces. So many are needed because each variety cuts the backbone of the DNA molecule at a separate, characteristic position in its complex sequence of chemical nucleotides. By using a series of different restriction enzymes, one can chop a stretch of DNA into segments with identifiable beginnings and ends, which may then be analyzed to determine just which individual chemical rungs (nucleotides) constitute each gene sequence.

Deciphering the actual order of the nucleotides is called sequencing; in a sense, it is the division of each genetic word into its individual letters. It is Thomas Hunt Morgan's once-revolutionary mapping of the fruit fly chromosomes carried to its most fundamental level. Before the advent of restriction enzymes, it could take as long as two years for a reseacher to identify even twenty of those molecular letters within a single gene. The new cleaving powers of the restriction enzymes have made it possible to do the same work in less than a day. "The discovery of the utility of the restriction enzymes," James Watson said in his Cold Spring Harbor office, "allowed us to work with material that no one, previously, would have dared to sequence."

In the months following Asilomar, sequencing seemed a more socially acceptable activity than using similar restriction enzymes to create recombinant DNA molecules. "I'd love to do recombinant DNA work," one young researcher at Cold Spring Harbor said. "But I've only got a year's fellowship here, and with the guidelines, that's about how long it would take to get that kind of experiment *started*."

But even the territory opened up by sequencing was broad indeed. Take, for example, the perplexing observation that there is really a great deal more DNA within the genome of any organism than is necessary to record the genetic information itself. Especially in the higher organisms, much of the DNA doesn't seem to code for any specific biochemical product; the number of genes actually needed to describe the creature seems far smaller than the total amount of DNA within each cell.

What is the rest of that DNA doing? Much of it, likely, controls gene expression—telling those so-called structural genes when to turn on and off. That's no small role—it's the reason, for instance, that one's whole finger is not made of fingernail—and it figures in everything from fetal development to the growth of tumors. It will also probably prove a major stumbling block for would-be genetic engineers: Even after the desired genes sequence had been identified and isolated, how can it be regulated and put to work?

The power of the restriction enzymes is obvious: They are chemical scalpels specifically designed for surgery on the DNA molecule. But is it possible following such delicate surgery to tell whether a test tube contains *just* those invisible gene fragments that are desired? And is there a way to separate one kind from another?

In a side room of one of the CSH labs there is a row of clear Plexiglas tanks, rather like the goldfish section of the five-and-ten, except that these tanks contain not fish but thin slabs of a substance called agarose gel. Each tank is connected by thick red and black leads to the terminals of a bank of variable power supplies. The tanks are used in gel electrophoresis—a formidable name for a fairly straight-forward process that separates large molecules of different sizes, such as segments of DNA that have been broken out of a particular genome chosen for analysis. The gel itself is a book-sized plate of translucent material an eighth of an inch thick with the look and feel of stale Jello. It hangs vertically within the plastic tank, and the mixture of fragments to be separated is placed in small indentations at the top edge. The power supply produces a current through the thin slab, inducing each fragment to begin a slow crawl down the length of the agarose, its speed determined by its size.

This process goes on for a period of time that ranges from an hour to a day, and then, when the current is shut off, the result is a gel through which fragments have spaced themselves according to size. A sharp scapel can be used to cut out the bit of gel containing the fragments one wishes to study, and when the gel is dissloved away, the result is a neatly sorted collection of specific segments of a DNA molecule.

What if one wants an assay more specific than just relative size—if one is on the trail of a particular sequence of DNA? This, most likely, is when one enters the room with radiation symbol on the door.

If one has a sample of the segment in question, one can, with some suitable biochemistry, produce what is known as a "complementary" sequence—a stretch that will find and bond irresistibly to its chemical cousins amidst even the most random mix of fragments.

These complementary sequences can be labeled beforehand with a radioactive isotope—carbon 14, say, or phosphorus 32. The "hot" sequences—often called "probes"—are mixed into the batch of random DNA fragments, given time to find their complements, and then the whole business is run through gel electrophoresis. One simply locks up the finished gel in a darkroom, flat up against a sheet of photographic film, and then develops the film. Each smudge on the emulsion will mark a congregation of DNA fragments that has taken up its radioactive complement, corresponding exactly to the fragments' position on the gel itself.

Cold Spring Harbor provided the basic answer to my question of how molecular geneticists manage their infinitesimal manipulations. In short, nonvisually. While molecular genetics is often popularly taught with colorful models of double helixes and minutely accurate electron micrographs of sinuous plasmid DNAs, at most points in the actual research scientists seem to have only a limited interest in what the subject of their experiment actually *looks* like. And so molecular genetics may well soon replace the romantic image of the biologist glued to his microscope with one showing a white-coated figure staring at bands on agarose gels or gazing fixedly at the readout of a scintillation counter.

But not all the tricks of molecular genetics are exclusively chemical. Likely the most powerful and controversial of the new techniques is biological indeed: the matter of molecular cloning.

Cloning, as mentioned earlier, is a process that has been accorded a sinister reputation by several generations of science fiction writers. It refers to the asexual duplication of anything from a single cell to an entire organism, based solely on the information contained in a single parent cell's DNA. It is the biological version of photocopy, and its most visible results thus far have been frogs and tobacco plants.

Its most interesting applications, however, have been in the realm of the genome itself. Suppose one has a short fragment of DNA that one would like to duplicate. One can, using the new recombinant DNA techniques, splice that fragment into a plasmid and then send that hybrid into a bacterium such as *E. coli*. A few bacterial generations—which reproduce the hybrid plasmid as well —result in a very large number of hybrid DNAs. After isolating those plasmids, one can use the same restriction enzyme that cleaved the plasmid in the first place to recover the original DNA fragment. And the result is an "amplified" gene or sequence of genes—a collection of perfect copies in sufficient quantity to allow intensive chemical analysis.

The technique is an interesting and profitable play on the border between life and chemistry. One uses the biology of bacteria to produce enough copies of a specific genetic sequence to allow one to drag the whole study back over into the territory of pure chemistry. It is a reminder of just how powerful biological processes harnessed by man can be: To build a gene sequence from raw chemicals, without nature's help, is only now even crudely possible—and still hideously complex and time-consuming.

Already researchers have started to amass libraries of cloned gene sequences, ranging from *E. coli* to yeast to the ribosomal DNA of frogs. The sequences—which, at this point, almost certainly contain more than just one genetic command—can be stored in containers as diverse as tiny vials in laboratory refrigerators, dots of dried chemical on cellulose filters, or minute wells formed in glass petri dishes.

The quickest way to build a library of cloned gene sequences is the controversial technique called shotgunning —breaking up the entire genome of an organism with restriction enzymes, plugging the individual fragments into plasmids or phage, and then growing all of them up within suitable bacterial hosts. Shotgunning was controversial even at Asilomar. Who could tell, went the argument, whether one of those random genes, introduced on a plasmid into something like *E. coli*, might not suddenly render that bacterium hazardous to humans or, for that matter, to some other unfortunate resident of the biosphere?

No one, of course, *could* tell—although almost everyone in the field had an opinion. But that uncertainty underscored just how much shotgunning offered in terms of previously inaccessible knowledge. With a collection of cloned genes one could, using radioactively labeled complements, begin to map with new assurance just where each gene and its control sequence resides in the chromosome. The cloned sequences could be introduced, one, by one, into bacteria genetically deficient in a specific biochemical function, and if the bacteria began to function correctly again, then one would know just what the newly implanted gene sequence coded for.

Sequencing and molecular cloning, along with the

broader potentials inherent in recombinant DNA techniques, clearly provided the first step toward the three prerequisites for any kind of true genetic engineering: a knowledge of which genes control which biochemical functions within a given creature, of how those genes are constructed, and of how they are turned on and off.

But even if all three of those concepts were fully understood, it would still be only a beginning. Take for example the often touted notion of using genetic engineering to cure diabetes—to introduce the gene sequences that control insulin production into the pancreatic cells of diabetic humans and thereby correct their incorrect enzymatic functioning.

The chances are excellent that by the time this book is published, someone will have already isolated and characterized the DNA sequence that codes for insulin production. But chances are also excellent that this work will be done for the purpose of implanting these genes into a prokaryotic organism like *E. coli* to produce pharmaceutical insulin. The final fix—transplanting that same genetic sequence directly into the cells of a diabetic—waits on a whole series of mindboggling technical difficulties. The most obvious problem is the development of a suitable vector, perhaps some mild human virus, to introduce the new genes into the existing chromosomes, and to integrate the new information, moreover, in such a fashion that it will actually function.

Gene therapy for diabetics is a relatively straightforward problem, all in all, compared to science-fiction notions like prenatal genetic manipulation. And so, the undeniable power of recombinant DNA notwithstanding, those who fear a brave new world just around the corner can probably rest easy, for the moment. Considering the substantial public outcry that the genetic manipulation of *E. coli* has already aroused, one can only imagine the storm that will surround the first researcher who proposes a recombinant DNA experiment involving an animal virus vector and human cells.

But that's not to suggest that both the products and problems of true genetic engineering are not already almost upon us. The academic laboratory is really only half of the experimental arena in which recombinant DNA techniques seem likely to make a rapid and significant contribution. The other half, of course, is industry.

During the months following Asilomar, rumors were rampant about projects underway at the major American and European pharmaceutical companies. Some were more than rumor. Great Britain's massive Imperial Chemical Industries announced, early on, a collaboration with the University of Edinburgh to investigate the potentials of recombinant DNA engineering and the feasibility of constructing a suitable containment facility for any potentially hazardous work. ICI's activities were closely followed by American pharmaceutical manufacturers, and while the strong tradition of proprietary secrecy in that business made such monitoring difficult, rumors circulated about notions ranging from newly constructed high-containment facilities to synthetic insulin genes that had already been developed in secret European laboratories.

Many of the most paranoid rumors seemed to originate, oddly enough, with academic molecular geneticists. Their special fear was that while the guidelines might contol the NIH-funded workers in the universities, industrial researchers would be free to perform whatever hazardous genetic manipulations they pleased. Those fears were compounded by the logical observation that the products that would most interest pharmaceutical companies—insulin, say, or growth hormones—would almost certainly involve the manipulation of higher mammalian genes, including human ones, which is just the kind of work that the NIH guidelines are intended to regulate most stringently. And thus the academics suspected that industry might upset the delicate self-regulation process—and while the strict federal legislation that would doubtless follow might be aimed at industry, it would inevitably also throttle the university scientist's research.

American industry reacted with suitable self-righteousness. Some representatives were even heard to offer the observation that if anyone got out of control, it would be in the university. "There's less freedom for the workers in industry to do anything they damn well please," one corporation president said. "You can't get a crackpot coming in at night and doing crazy experiments, which might be reasonably easy in a university environment. There are more people in industry to blow the whistle, and there's more fear of liability actions."

That last may be the telling point. At any rate, American industry snapped to with remarkable alacrity. Representatives of several major pharmaceutical manufacturers were in conspicuous, though silent, attendance at the early NIH guidelines committee meetings. And just three weeks before the issuance of the final guidelines, an industry-wide meeting at the NIH campus in Bethesda, Maryland, managed to draw more than twenty-five representatives from such corporate giants as Eli Lilly, Upjohn, General Electric, Monsanto Chemical, and Union Carbide.

The thrust of that meeting seemed to be that industry, recognizing its potentially disruptive role outside the purview of the NIH guidelines, would take the part of the parson's wife and remain scrupulously above criticism. At the same time they strongly preferred that the guidelines remain voluntary. Two provisions of the final NIH guidelines, in particular, did not sit well at all with anyone who had an eye on commercial applications. One put stringent limitations on work with volumes of culture solution greater than ten liters; the other required investigators to inform NIH of new safe host/vector systems and to make those strains available to other researchers.

The former provision, by the time it reached the final guidelines, had somehow acquired a clause exempting experiments "of direct societal benefit," which was without question a large enough loophole to drive a pharmaceutical company through. The spirit of the latter requirement, however, was more of a problem, flying spectacularly, in fact, in the face of the long-time tradition of deep secrecy that has characterezed the drug companies ever since the discovery (and mass production) of antibiotics.

Antibiotics are, in fact, a good example of just how recombinant DNA could alter the business of exploiting microorganisms. Ever since Sir Alexander Fleming's fortuitous discovery, in 1928, of a fungus that excreted a bacteria-killing substance he called pennicillin, the search has continued unabated for new strains of microorganisms that would either produce larger quantities of known antibiotics or else make new ones altogether. The most successful of the early strains of *Penicillium* was discovered, for example, on a moldy cantaloupe in a Peoria, Illinois, fruit market. Equally intense searches, involving soil samples harvested from every corner of the planet, preceded the discovery of two other early antibiotics, streptomycin and aureomycin.

While human beings have prompted microbes to do everything from producing acetone and medical steroids to consuming industrial waste, the fact remains that microorganisms basically only do what their genes tell them to. There was a saying in the industry that "the bug of discovery is the bug of production"; in other words, one is fairly well stuck with the inherent genetic abilities of whatever useful microorganism one has managed to find. That limitation hasn't been completely insurmountable, even without the new recombinant engineering techniques. One common practice has been to mutate genes with doses of chemicals or radiation, and by now drug companies spend considerable time barraging microbial genomes with a variety of mutagens in hopes that some subsequent mutation will increase the creature's antibiotic yield. Of course, it's a long shot; most of the time, the induced mutation kills the microbe outright.

It's not difficult, then, to see the immediate usefulness of recombinant DNA for the drug industry. If the gene sequence for a given antibiotic could be isolated and decoded, the information could be utilized to design and produce a microorganism that would manufacture that antibiotic at high efficiency and low cost. The same is true for animal proteins like insulin, which must still be harvested from costly animal tissue; microorganisms don't make insulin, simply because they don't need it. And thus this is precisely the arena in which the more farsighted proponents of recombinant DNA see the greatest immediate benefits—as in Joshua Lederberg's prediction that someday microorganisms might well be programmed to produce such sophisticated human proteins as antibody globulins.

It's an undeniably exciting prospect—sufficiently exciting, in fact, to spawn a handful of new corporations and attract the unwavering attention of more than a few of the *Fortune* 500. And while it remains to be seen just how industry will ultimately deal with the NIH guidelines (unless some abrupt congressional action removes that option altogether), it is already clear that the future of commercial synthetic biology will offer some very perplexing questions. One of the most interesting, in fact, sounds like science fiction: Can one actually patent a new life form?

The question is more than simply speculative. It came up initially in Schenectady, New York, which calls itself "The Electric City," in deference to the massive General Electric manufacturing and research complex that dominates the town's economy. While not the largest industrial laboratory in the world, GE's Research and Development Center, perched over the banks of the Mohawk River, is exceedingly active: In 1975 GE gained a patent a day, on the average, from the Schenectady R&D Center. One of those patents, however, was rejected that year, a decision that was immediately appealed.

The patent application had involved a genus of bacteria called *Pseudomonas*, a microbe found in soil and water, and which displayed, on occasion, an appetite for crude oil. Some even considered the presence of *Pseudomonas* a potential indication of underground petroleum. A young GE researcher named Ananda Chakrabarty decided to investigate the possibility of recruiting a *Pseudomonas* strain to clean up oil spills at sea. But unfortunately, he found that no single strain existed in nature that could digest all of the complex hydrocarbons that constitute crude oil.

Chakrabarty, however, knew that plasmids were the genetic factor that determined whether a *Pseudomonas* could digest a given component of crude oil. And so he developed a laboratory technique to introduce a whole complement of varied plasmids into a single *Pseudomonas*. The plasmids would not have tolerated each other in the same bacterial cell in nature; Chakrabarty's accomplishment was to convince the plasmids that they could, in fact, live together. The result was, in effect, a new strain of *Pseudomonas* that was a super oil-digester. Chakrabarty's experimental success led GE to contemplate packaging a dried inoculum of the bacteria, say, coated onto the dry straw that is customarily dropped on large ocean oil spills.

Shortly thereafter, the press started regularly referring to "oil-spill removal" as one of the purported benefits of recombinant DNA technology; one imaginative critic even developed a scenario wherein Chakrabarty's *Pseudomonas* invaded the lubrication system of a 747, with disastrous results. These, of course, were based on misconceptions. Bacteria cannot live in pure oil; and neither did the GE work really involve recombinant DNA engineering in the first place. DNA molecules had neither been cleaved nor rejoined, since Chakrabarty had been deft enough to find natural DNA sequences, in the form of plasmids, that already coded for the hydrocarbon digestion he wanted. But the distinction didn't seem to make much difference in terms of publicity. And in the end, it didn't make much difference either, in terms of the lessons that the GE experience offered for the future of commercial genetic engineering.

Two problems developed immediately. The first was a question of environmental impact: What disturbance might result from dumping a new, genetically engineered microorganism into the ocean? And how do you find the answer to this question without actually doing it?

The Environmental Protection Agency—which has, in recent years, evaluated several proposals to use natural bacteria for oil-spill removal—had no guidelines past the requirement that the oil-eating culture could not contain certain microorganisms pathogenic to man. What about long-term environmental effects? "That's a good question," one EPA researcher said, late in 1975, "and based on our current procedures, it wouldn't be asked." As far as NIH-funded work goes, however, that concern is effectively covered: The final guidelines specifically prohibit "the deliberate release into the environment of any organism containing a recombinant DNA molecule." And the GE work has received sufficient publicity so that one suspects it will require some fairly persuasive evidence from the manufacturer in order to convince local environmental watchdogs that a genetically engineered bacterium is the sort of microbe that one would want in one's own harbor.

The second problem encountered by the GE *Pseudomonas* project was the science-fiction-like question presented earlier. In the spring of 1974, GE was granted a patent on the complex laboratory process that Chakrabarty used to fit the wild-type *Pseudomonas* with multiple plasmids, but at the same time, not for the product of that process—the modified bug itself.

The latter patent application, as of late 1976, remains on appeal and its immediate future is uncertain. Its import, however, is clear: Sooner or later some decision must be made as to whether one can patent a new living organism. The question, almost certainly, will only grow more pressing.

The precedents in the matter are foggy. One can patent hybrid plants, like roses and fruit trees, so long as they are reproduced vegetatively, i.e., grown from rootstock. Plants grown from seeds, involving the uncertainties of cross-pollination and such, are considered too inherently unstable to patent. There have been, in fact, unsuccessful attempts in the past to argue that bacteria *are* plants and should be provided equivalent patent protection. At present, about the best a firm can do to protect its proprietary bug—such as a bacterial strain that is unusually efficient at antibiotic production—is to patent the whole process in which the microorganism takes part and then keep the microorganism itself carefully locked up.

This leads, of course, to a considerable level of paranoia, particularly in the pharmaceutical industry, where elaborate security systems and deep secrecy surround valuable production bugs, a situation altogether antithetical to the open atmosphere that NIH hopes to foster for the future of recombinant DNA work.

Whether or not GE carries through its fight to patent its *Pseudomonas,* the questions, both legal and environmental, created by the commercial potentials of genetic engineering cannot possibly go long without answers. GE's Chakrabarty, for example, was already considering modified microorganisms capable of harvesting heavy metals like gold or platinum from waste substances like the insides of discarded automobile catalytic converters. The rumors about the pharmaceutical industries continue, and it seems only a matter of time before the era of billion-dollar bugs—possibly capable of generating exotic products difficult now to even imagine—will be upon us.

It will demand caution, at this point, to draft legal precedents establishing the right of an individual to own sole rights to a self-reproducing new life form, and yet that seems almost an inevitable step. It is an abruptly practical question about the future of synthetic biology that for the moment sounds almost as much like science fiction as my old fantasy about the synthetic muscle. But the chances are very good that these strange new dilemmas will stay with us long enough that, in time, they will no longer seem strange at all.

11

The Bug Disarmed?

Early in December 1975, ten months after Asilomar, I checked into a plush hotel in La Jolla, California, a lavish resort community just north of San Diego that was founded on the traditions of golf, tennis, and exclusive ocean frontage. These days, it has become less exclusive, playing host to the free-spending hordes that regularly stream south from Los Angeles, but some of the old ambience still hangs on, cloistered and distilled in, among other places, the Spanish-style hotel at which I was staying, the La Valencia.

That gray week in December, the La Valencia was the scene of what was then optimistically billed as the final guidelines meeting of the NIH Recombinant DNA Molecules Advisory Committee—those fifteen people charged with writing the specific guidelines for the new technology.

The fifteen were meeting at nine in the morning on the fourth of December, and as I knotted my tie in my room an hour earlier, I may well have been recalling the turbulence that the committee had already generated during its few previous meetings. Right now, all I really remember is the big color television and the religious talk show that the local station broadcast at that hour. The subject that morning was the theory of evolution. Two fundamentalist Christians were the guests: one young, plump, and unctuous, the other old, plump, and pugnacious.

Neither, predictably, had much sympathy for the "theory." The younger referred repeatedly to our supposed origins in the "primordial goo." The elder sneered outright. "There's a fellow at Harvard," he confided, "who believes in *quantum evolution!* He believes that one day a bird sat down and laid an egg and a whole different animal came out! That's a straight-faced egghead theory," the plump gentleman declared, grinning right into the camera. As host and guests guffawed in unison, I switched the set off and headed out into the hall. A few hours later, however, in the meeting room four floors below, I could not help recalling their guffaws.

The discussion around the T-shaped conference table was just then heading into the sticky matter of whether human DNA should be treated with more laboratory caution than DNA derived from the lower primates. In the midst of confusing—and occasionally confused—technical exchanges, I thought back to the TV program and realized that, all in all, the plump old evangelist had really pinned down the plushest niche in the evolution racket. The loyal opposition at least offers steady work. The avant garde—even at that early juncture—clearly meant nothing if not the worst kind of headaches.

Several hundred pages of those headaches were already arrayed before the committee members in the La Valencia meeting room—in bright red notebooks containing an assortment of the letters that DeWitt Stetten, the committee's chairman, had received in the months following their previous meeting at Woods Hole, Massachusetts.

Stetten had not lacked for correspondents during those intervening months. And most of the letters were more or less incensed observations—often from researchers who had attended Asilomar—that the committee had sold the Asilomar statement down the river by approving, at Woods Hole, a diluted set of guidelines. While a handful of letters argued that the guidelines were excessively strict and would throttle research, the consensus was clearly that the Woods Hole guidelines had significantly reduced the levels of containment and caution—especially with regard to shotgunning experiments—below that suggested by the Asilomar statement.

The strongest reaction of all was a petition, addressed to Stetten, and signed by forty-nine biologists who had attended a bacteriophage workshop at Cold Spring Harbor, just one month after the Woods Hole meeting. "We are concerned," the petition began, "that the present draft appears to lower substantially the safety standards set and accepted by the scientific community . . . at Asilomar in February, 1975." That petition was the first real organized criticism of the guidelines, and it attracted immediate media attention. Headlines ranged from *Science*'s "NIH GROUP STIRS STORM BY DRAFTING LAXER RULES" to the *Washington Post*'s more terse "GENETIC ENGINEERING DIVIDES SCIENTISTS."

By the time the La Jolla meeting convened, four months after Woods Hole, not only were scientists divided, but so were the guidelines. By then, three different versions existed, all drafted over the previous nine months and none of them was acceptable to everyone. The differences between versions ranged from substantial technical issues to the line-by-line placement of semicolons. The central questions, however, were similar to those that had been debated the most at Asilomar: Was *E. coli* the best vessel for these first tentative genetic manipulations? What unknowns, viral or otherwise, lurk within the unexplored territory of the genomes of higher organisms? How safe is it to explore such unknown country with a technique so random that even its proponents call it shotgunning? What kind of laboratory precautions—both physical and biological—should the new work require? And, in the end, can even the most stringent laboratory precautions ever prove fail-safe?

The audience at La Jolla included representatives of organizations as diverse as the Environmental Protection Agency, NASA, *Time*, Eli Lilly, and the MIT Oral History Archives. There was a palpable tension in the air; one could sense that the recombinant DNA question was rapidly dividing into an adversary situation. On one side, there were those who insisted that some of the most valuable biology of the century had already been unconscionably delayed, and on the other, those (particularly among the younger researchers) with a new sense, only hinted at by the Cold Spring Harbor petition, that the whole notion of gene reshuffling should be reevaluated from the outset. It was time to come up with some generally acceptable guidelines, and without delay. The renewed press attention, the barrage of letters, the Cold Spring Harbor petition were all unsubtle reminders that matters might soon get out of hand—and into what other hands, no one really wanted to speculate.

Many researchers still recalled Joshua Lederberg's stern warning, that first day at Asilomar, about the likelihood of the recommendations crystallizing into legislation. By the time of the La Jolla meeting, that possibility seemed closer indeed, due primarily to the efforts of Massachusetts Senator Edward M. Kennedy.

Kennedy, the chairman of the Senate Subcommittee on Health, had previously been considered somewhat of an ally of scientists against excessive political pressure. He took an early interest in the recombinant DNA issue, however, and toward the end of April 1975, he and his subcommittee spent a day questioning a panel of four individuals about the guidelines and the way they had been decided on. His principal criticism centered on the omission of public representatives from the decision-making process, and his concern at first seemed less the dangers of recombinant DNA than the potentially unhealthy model that Asilomar and the guidelines process might represent. But much of his questioning involved technical details, and most of that was directed at Stanley Cohen, the young Stanford researcher who had done some of the earliest recombinant DNA work, but who was also slightly reticent about speaking out publicly on the matter.

Kennedy, however, brooked no reticence. "Are these issues too complex for lay people to understand?" Kennedy asked, and Cohen was unhesitant: "Not at all, Senator." Cohen proceeded to deliver one of the more coherent summaries of recombinant DNA work that the senator was likely to hear, emphasizing his own work with plasmids.

But Kennedy finally broke in. "Have there been," he asked somberly, "any accidents, or have there been any problems in controlling the material?"

"In controlling *which* material?" Cohen wondered.

"The plasmids," said Kennedy.

Cohen took a deep breath. "I am not aware of any accidents, or consequences of any."

Kennedy pressed the point. "There has not been any diversion of the material at all?"

Now Cohen looked distinctly puzzled. "Any what, Senator?"

"Diversion," Kennedy repeated carefully. "Has anybody gained access to this material who should not have? Has there been any that has been misplaced or removed or taken that you know about?"

"The material?" Cohen asked tentatively. "Do you mean bacterial plasmids?"

Kennedy mulled the question for a moment and then retreated. "What is the problem?" he asked finally, contritely. "I'm a layman and new to this. Has there been any material that has been either taken, removed, that has not been accounted for?"

Cohen stared down at his notes. "Plasmids," he said tersely, "are naturally occurring genetic elements that are in many bacteria found in nature. Since bacteria and plasmids are self-propagating, they can't be considered in terms that imply a specific quantity."

Kennedy abruptly changed the subject—and thereafter his staff, presumably, stopped analyzing the issue quite so specifically in terms of nuclear energy. Less than a month after the hearing, however, in an address at the Harvard School of Public Health, Kennedy launched a direct attack on the "inadequate" Asilomar conference: "Scientists alone decided to impose the moratorium, and scientists alone decided to lift it."

David Baltimore, at nearby MIT, was bewildered by the speech. "Kennedy misunderstood the whole process," he said. "We weren't arguing that Asilomar was the last word—but at that time, we were the only ones who knew what was going on, and our whole point was to alert the public."

The public was, for better or worse, alerted. And so, by the time the Recombinant DNA Molecules Advisory Committee reconvened at La Jolla, public opinion had itself become a real issue. Kennedy's pronouncements had already created mild paranoia among members of the committee. Lederberg's warning might well prove accurate, and the eventuality of legislative intervention was one that no one relished. Perhaps partly as a result of that, a number of influential researchers not on the committee itself showed up at La Jolla, including Paul Berg, Sydney Brenner, Maxine Singer, and Stanley Cohen.

Despite the pressures of time and public scrutiny, the La Jolla meeting proceeded with a greater sense of order and harmony than had Asilomar. But then at least some of the impetus for the final decisions came from the reassurance of another meeting, convened just four days earlier—a meeting that had been as relentlessly technical as La Jolla was political.

Few reporters attended the "Workshop on Design and Testing of Safer Prokaryotic Vehicles and Bacterial Hosts for Research on Recombinant DNA Molecules," which was held just twenty miles up the rocky California coast from La Jolla at an informal golf resort called Torrey Pines. The workshop was the first stock-taking of Sydney Brenner's urgings at Asilomar toward "disarming the bug," or, creating biologically safe host/vector systems. That curious inversion of evolution—the intentional construction of life-forms that cannot survive outside the laboratory—had required far more effort than even Brenner's critics had imagined.

Just after Asilomar, at the first NIH committee meeting in San Franciso, the members agreed that Brenner's "Mach II" optimism was excessive and that no one could expect a safe, disarmed *E. coli* vector in less than three months. Nine months later, the Torry Pines workshop convened to determine just how much longer than that it might take. Nature, it seemed, would not be disarmed without a struggle. "In terms of back up systems" one researcher observed at Torrey Pines, "*E. coli* makes NASA look sick."

Yet the concept of "disarming the bug" was a central part of the effort to make recombinant DNA work socially acceptable. And until safer host/vector systems were developed in effect, the moratorium continued on all work that required their use. If the technology was going to take so revolutionary a step forward, it seemed fair that the means of containment should do the same. And almost without exception, no one who had firsthand experience with the casual nature of traditional laboratory containment procedures wanted to depend on them.

Paul Berg described those fears in a letter to Stetten just following the Woods Hole guidelines. "I'm convinced," he wrote, "that physical containment is overrated and, while reassuring to the psyche, is hardly the line of defense one would like to put the greatest reliance upon."

Berg described some of the difficulties he'd encountered in his own costly P3 laboratory. Many of his workers, he noted, even at that containment level, had developed antibodies in their blood that indicated they had been infected by the SV40 particles with which they worked.

Berg provided a list of potential accidents (careless mouth pipetting, punctured fingers, solutions poured down drains, centrifuge explosions, spills on the bench) and then concluded that "the ideal safeguard is ingeniously designed biological containment that can prevent escape and propagation—even with slobs doing the work on open benches."

Such ingenious design was precisely the point of Torrey Pines: The sixty-two researchers who gathered in the California meeting room represented one of the more deliberate attempts at genetic engineering thus far—the intentional laboratory creation of the unfit. The seating, acres of green golf course at the edge of the Pacific, was not without irony. Past the picture window, dotting the golf course, were occasional stands of the all-but-extinct Torrey Pine—a small, gnarled tree that had been driven back by the last Ice Age to this single, final toehold on the coast of California. A survivor indeed. And within the lodge that bore its name, from the very first presentation on, proceeded a relentless, genteel, and esoteric contest as to precisely who had managed to design the most unfit microbe of all.

Would the newly disarmed vehicle be a crippled bacteriophage? An extensively mutated plasmid? For host duties, how about a massively altered *E. coli* that stumbles through its nutrient medium like a leaking gas bag, sustained only by the biochemical largess of its principal investigator? Or maybe not *E. coli* at all. Perhaps a *Pseudomonas,* or *Bacillus subtilis,* or even the obscure *Megaterium.*

The approaches were varied and ingenious: complex genetic manipulations to create, say, bacteria whose cell walls would collapse like punctured cellophane if not maintained in a medium containing a specific chemical component that the creature was constitutionally unable to make for itself; or bacteriophage and plasmids unable to function or reproduce unless introduced into specific bacteria which were, themselves, crippled—microorganisms, in short, with so many built-in self-destruct factors that it would take a miracle to sustain them outside their laboratory glassware.

That was the idea, at least. The question at Torrey Pines was who could manage that debilitation best. It was already well-known that NIH had put out requests for grant proposals in the area of modified host/vector systems. It was also well-known that just three days later and a few miles down the coast, some fairly binding decisions about the future of recombinant DNA would be made. And thus there was more than a bit of an edge to the seemingly benign technical competition at Torrey Pines.

What *is* the best way to cripple a microorganism selectively? By the time the Torrey Pines workshop was held, more than a few researchers wished they hadn't asked that question in the first place. One of them was Roy Curtiss, the lanky, long-haired researcher from the University of Alabama who was by then probably the leading researcher in the crippled-host field. Just after Asilomar he'd started work on the familiar and dependable *E. coli*; and in the months following, using a whole spectrum of exotic techniques, turned that ubiquitous bacterium into a guaranteed loser.

Curtiss had already logged more than twelve years of laboratory experience with *E. coli*. His final assault on the bug involved nine months of work, with the assistance of twelve coworkers. By Torrey Pines, that added up to several thousand laboratory man-hours—and even then, Curtiss could not claim complete success.

Curtiss had, ironically, been one of the first voices to speak on the side of the inevitable mutability of nature. After the elaborate English presentations on the ecology of *E. coli* K-12 at Asilomar, Curtiss had immediately suggested that while the survival rate of that bug in the human gut was important, the bacteria's survival in, for instance, sewer systems might well be equally pertinent.

Following Asilomar, Curtiss sat on the NIH Recombinant DNA Molecules Advisory Committee, where one of his favorite observations, whenever the discussion of potential hazards grew excessively anthropocentric, was the reminder that "we depend on other organisms, and hardly any of them depend on us. If they had a mind, they'd probably just as soon do without us."

That concern found a strange expression at Torrey Pines. A continuing dispute, of course, whether *E. coli* should be chucked altogether as a potential host, in favor of some bacterium that was known not to colonize humans. The dispute was made more complicated by the fact that, soon after Asilomar, both the press and many critics were consistently confusing *E. coli* K-12—which possesses little apparent ability for infecting humans—with its far more infectious wild-type brethren. Whether because of *E. coli*'s bad press or its inherent drawbacks, many scientists at Torrey Pines offered suggestions for alternative hosts. GE's Chakrabarty, for example, nominated his stable of *Pseudomonas*. Another researcher gave a pitch for *Bacillus subtilis* as a candidate, and after a few moments he peeled off his workshirt to reveal a bright yellow T-shirt neatly imprinted with the chromosome map of his favorite microbe.

B. subtilis, in fact—while rather a genetic unknown compared to the extensively studied *E. coli*—seemed to have some real credentials for alternative host duties. Most importantly, it disdains the human digestive tract, vastly preferring materials like plain dirt or rotten hay. While some members of the audience were quick to point out that certain strains of *Pseudomonas* are known to be opportunistic hospital pests, no one could seem to find a similar strike against *B. subtilis.* It likes plain dirt, the speaker emphasized—not *Homo sapiens.*

There was brief silence in the Torrey Pines conference room, and then, one by one, gazes wandered out through the broad picture windows overlooking the carefully tended golf greens. "But what," a voice near the back of the room asked at last, "about the *golf course?*"

What indeed? It began to look as if, regardless of what host one proposed, there was something out there —the grass, the soil microorganisms, the indigenous invertebrates, the caddies—that was fair game for some kind of imaginative disaster scenario. Several months afterward, some researchers would go so far as to propose the use of thermophilic bacteria—rare and fairly obscure species that prefer to live only in hot springs— theorizing that that would limit the risk.

Curtiss's experience with *E. coli* K-12 underlined how abruptly bacterial genetics was being pushed to its limits. One was, after all, bucking several billion years of evolution, and our understanding of *E. coli*—probably the most extensive in microbial genetics—suddenly seemed insufficient when one actually had to assume responsibility for the bacterium's behavior in the world.

The microorganisms simply weren't cooperating. "You have something one evening," someone at Torrey Pines said, "and then it's gone in the morning." It seemed axiomatic that if something could go wrong with a modified host, it would. Nature's intransigence was making it difficult for humans to keep their first tentative genetic dabblings safely within the laboratory.

Torrey Pines brought home the fact that, in a broad sense, the recombinant DNA controversy represented molecular geneticists' first real dues-paying for the structural ubiquity of DNA in nature. It had been that ubiquity—the striking similarity between the means of genetic coding for all creatures—be they bacteriophage or human being—that had, in previous decades, allowed the fledgling science to shift its researches so easily from fruit flies to bread mold to *E. coli* and phage, and thus rapidly describe some very basic genetic mechanisms.

Suddenly, however, that freedom had come home to roost. If DNA sequences were in fact so interchangeable, might a random bit of, say, toad DNA begin to reprogram, in some altogether unforeseen fashion, a previously benign soil bacterium? What if a hybrid DNA molecule managed to undo all the work that had gone into disarming its host cell in the first place?

No one at Torrey Pines—or anyplace else—really knew. The same fundamental rationale that had delivered molecular genetics at express-train speed to its present sophistication could easily be turned around to justify unbounded paranoia about the future of synthetic biology. It could, and in the months to come, it would.

But perhaps biology would in fact not prove nearly so susceptible to human intervention as researchers might flatter themselves to think. If anything, the difficulties at Torrey Pines tended to fall more along those lines.

The maps of modified genomes—bacteria, plasmids, phage—went up on the blackboards and projection screens like the blueprints of next year's models at a stockholders' meeting in Detroit. Most had something wrong with them, and the rest of the audience was more than willing to take their best shot at deflating the proposal. "What about propagation in sewage? Do you have figures?" "Aren't bacterial cell surfaces environment-dependent? What if phage resistance isn't exclusively genotypic?" "Isn't *B. subtilis* a problem in pacemaker implants?" "Is antibiotic resistance really acceptable as a marker?" "The world, remember, is not a flask."

The room contained some of the sharpest minds in molecular genetics, and the competition was keen. "Someone," one observer said drily, "always has to go to the head of the class." It was serious business, and while humor was welcomed, speaking through one's hat was not. At one point, for example, one young crewcut California researcher, apparently on the spur of the moment, stood up to describe his own notion of a disarmed vector. It was an odd-sounding plasmid, for which the young man offered a vague, sketchy description of how one might construct it. "A few tricks, here and there," he summed up. By the end of this extemporaneous presentation, much of the audience was staring rather coldly. A Cold Spring Harbor rseearcher asked if the man was going to make this plasmid himself. The researcher smiled and shrugged amiably. "I don't know," he said. Then there followed a pointed silence until the young man sat down again.

By the end of the Torrey Pines sessions, it was clear that Roy Curtiss had managed the most remarkable advances in terms of modified bacteria. His version of *E. coli* K-12 was loaded down with an impressive set of mutations: Besides being burdened with highly specific biochemical needs, Curtiss's bug was designed to self-destruct at the temperatures it would encounter within mammals, and, moreover, it was unable to transfer DNA in that curious bacterial sex act called conjugation. It wouldn't even allow much truck with bacteriophages—which can, under certain circumstances, act to transfer genes between neighboring bacteria. The sum of those plus additional, fail-safe mechanisms seemed to have produced a bacterium whose chances of surviving outside a highly specific laboratory environment were around one in a billion.

That was approximately the survival figure that the NIH committee had suggested a modified host would need to qualify for EK2 status. It would still be months,

of course, before the NIH recommendations were final-
ized—and months again before Curtiss's microbe would
become the first EK2 host to win approval.

Torrey Pines recalled, once again, Robert Oppenheimer's disclaimer during the Manhattan Project:
"There are no experts," he would tell outsiders. "The
field is too new." After one morning's presentations by
specialists on bacteriophage, I noticed three leading
plasmid researchers chuckling together over in one corner of the meeting room.

What, I wondered, was so funny?

"These phage people," one of them answered after a
moment. "You can't understand a word they say."

There was, of course, some tongue-in-cheek there—
although it did seem a bit disconcerting that researchers
who worked with the phage lambda were starting to call
themselves lambdologists. And a few days later, at La
Jolla, one of the NIH administrators who had been involved with the guidelines procedure from the outset,
confided with some resignation that he figured only
about one hundred people on the whole planet were
really fluent in this new business to begin with.

One who clearly was fluent was Sydney Brenner, who
arrived from England toward the end of the Torrey
Pines meeting. He brought word that English thinking
tended to favor the use of disabled phage, rather than
plasmids, as safe vectors. And he also brought some interesting observations on the whole notion of elaborate
physical containment.

Brenner thought that the grand and costly negative-
pressure facilities that Americans had been touting in
the months since Asilomar might not be the way to go.
To illustrate the point he told a story about a physical
chemist he once knew who had been doing work on the
degradation of propane at 750°F and extremely low
pressures. "On the plane back from Asilomar," Brenner
said, "it occurred to me that if *we* were doing that kind
of work, we'd be trying to figure out how to do it in a
lab maintained at 750° and a tenth of an atmosphere—
without harming the researcher."

Brenner's suggestion was that it made more sense to miniaturize experiments than to build vast facilities to contain them. Perhaps, he said, we can even learn to design experiments that minimize sonification and centrifugation—the two techniques biohazard experts consider most likely to spread contaminants. Brenner quickly ran through several examples of how, in given experiments, such "miniaturization" could take place. His examples were predictably shrewd and simple—but then not everyone in the field was ever going to be able to design experiments as deftly as Brenner.

He raised another potential problem, as well. "It seems to me," he said, "that many different host/vector systems will soon be required, for many different purposes." Does NIH, he wondered, have enough money to follow through on the potentially expensive certification procedures such multiple vectors might require?

The day before, a young American researcher had suggested that the responsibility for checking each safe host/vector system should rest with the researcher who wanted to use it in the first place. "If you can't do the tests," he said, "you can't clone—believe me."

Brenner concluded his observations at Torrey Pines that afternoon with a single terse comment about the responsibility of researchers during the first tentative work with recombinant DNA: "The real point," he said, "is simply not to inject into nature anything that will ultimately confer additional selective advantage."

It was an apt and succinct stricture, but it was also far easier said than done. The issue of human intervention into the evolution of microorganisms posed questions that would, a day or so later, in the midst of the La Jolla discussion, ultimately seem little more tractable than the problems of disarming *E. coli.*

12

The Scientists
Draw Their Lines

The NIH Recombinant DNA Molecules Advisory Committee meeting at La Jolla opened on Thursday morning with a brief observation by chairman Stetten: "It is, I hope, the feeling of everybody around the table that by the time we break up on Friday afternoon, we should have some sort of document prepared."

While the wood-paneled library of the La Valencia, decorated with murals, brass sconces, and elaborate chandeliers, seemed isolated and exclusive, there was little question that the outside world—from Senator Kennedy to the Cold Spring Harbor petitioners—was beginning to crowd around. A few days earlier, a front-page story in the *Washington Post* had dubbed La Jolla "the Scientific Showdown at the OK Corral."

Months later, the NIH committee would be accused by some of steadfastly ignoring public opinion. Yet during those days at La Jolla, the more astute of the committee members appeared almost as concerned about the reaction of the public as that of fellow researchers. Colleagues could at least be reasoned with, while the public was an unknown quantity altogether.

By the time it convened at La Jolla, the committee had, moreover, backed itself into a remarkable procedural corner: The guidelines existed in three different versions, varying from the insertion of several new paragraphs to the deletion of single semicolons. Within one line that described a particular laboratory procedure, version A might use the words "sharp, pointed"; B, "sharp and pointed"; and C, simply "pointed." And there were, at best count, more than 246 of these tripartite variations, not counting a whole blizzard of individual inserts—thus creating, among other difficulties, a nightmare in photocopies.

Outside the meeting room, a predominately geriatric population dozed in the ornate hotel lobby, amid overstuffed furniture and tables topped with half-completed jigsaw puzzles. Through the windows, the sky appeared overcast; the California sun had not yet broken through the morning fog. Within, the NIH meeting commenced briskly. Chairman Stetten suggested carefully that the first one hundred less controversial items of the guidelines might be accepted by general acclamation, unless anyone had specific objections. The specific objections ensued immediately.

This wasn't, it became obvious, going to be easy, even though most of the NIH committee members were old hands indeed. There was, however, one notable new inclusion: Elizabeth Kutter, a Ph.D. in biophysics, from Evergreen State College in Olympia, Washington. Kutter had been brought in, after the unhappy reaction to Woods Hole, to write the third set of guidelines, on the rationale that she had had little to do with the business from the outset.

Kutter was young, energetic, somewhat counterculturish, and exceedingly enthusiastic. She willingly described how a whole mélange of students had helped draft the new version, and thus she was mildly mistrusted from the outset. *Science* magazine suggested the prevailing attitude toward her with its characterization using a subtle parody of a popular television commercial for an orange drink.

Yet while no one knew quite what to make of Kutter, she proved to be just what the thoroughly divided group needed: someone that no one could really trust. One of the other female committee members took an immediate dislike to Kutter and frequently responded to the young woman's comments with muttered evaluations that ranged from "bad idea" to simply "silly, silly."

Kutter set the controversial tone of La Jolla early on. In a discussion of whether HEPA filters—the High Efficiency Particle Accumulators that were already standard equipment in the most rigorous biohazard facilities—should be installed as a matter of course in many recombinant DNA labs, Kutter mentioned in passing that such filters are really not that expensive. Another committee member, a department chairman at Yale, balked immediately, clearly puzzled. He'd just priced one for his own lab, and they cost in the neighborhood of a thousand dollars each. That's cheap? Someone else wondered just how much it would cost, in terms of funds available for real research, if a thousand labs across the country were required to install the HEPA filters. It would cost too much, another committee member answered, considering the minimal risks involved in much of the less controversial recombinant DNA work. But, Kutter asked, how much risk is worth how much cost? The argument was on.

Should unfiltered laboratory air ever be allowed to reach the environment? The sole layperson advising the committee—a law professor from Texas—allowed that this didn't sound like such a great idea. Stetten injected a humorous epigram: "Until the EPA made us honest, we used to say that the solution to pollution was dilution." After permitting considerable discussion, Stetten finally asked Emmet Barkeley—the National Cancer Institute head of biohazard control—for his opinion. "Experience teaches us," Barkeley said, with customary caution, "that general exhaust air does not produce insult to the environment." He added that the contamination of laboratory personnel, who then carry the undesirable bug out into the world, is really a far greater hazard.

"Wait a minute," said Paul Berg, who was sitting off to one side of the big conference table. Berg was no longer a member of the committee; both he and Maxine Singer had, understandably, declined reappointment, yet both found it difficult to remain uninvolved. "Wait a minute," Berg said. "This could be construed to allow P3 work in unfiltered rooms without negative pressure."

The small plastic egg-timer that Stetten was using to limit discussions began to ring. Outside the windows, the sun had long since reached the beach, yet the committee had managed to reach only line 59 of the 244-line composite edition of the guidelines—and the first part was universally considered the easy part. Stetten seized his egg-timer. "I'm afraid," he said, "that we're already behind schedule."

The issue of the HEPA filters was decided relatively quickly, with a bias toward the more conservative viewpoints—the same cautious direction in which most of the two days' controversies would be resolved. Less than twenty lines later, however, the process was held up again over a blanket provision forbidding the release of recombinant DNA molecule-bearing organisms into the environment. "Would this also cover new nitrogen-fixing crops?" someone asked. Kutter said yes, and there was a prompt motion to drop the whole clause, lest it discourage critical work in agriculture.

Peter Day—a botanist from Connecticut who had been brought in by the committee when it grew distressingly clear that none of the original members had much knowledge of that field—tended to agree. Agriculturists, he said, have a different definition of the environment than other researchers; but if the guidelines simply made it clear that newly created plant species in controlled greenhouses were acceptable, then there would be no problem. It was a touchy point, and after some discussion—"But there *are* no *controlled* field tests!"—it was decided, on a 6 to 7 vote, that a new version of the line should be drafted to express that greenhouses were not covered by the restriction.

Three hours later, the discussion had managed to reach line 100. Stetten, who suspected that the discussion would reach "a colossal halt" at line 110—which involved a rigorous new set of criteria for EK2 vectors—suggested that he might read quickly through the intervening sections and request blanket approval. And he did so, interrupted only by one committee member who regularly demanded, "Where are we? Where are we?"

Just before the discussion attempted to tackle the controversial matter of certifying safe host/vector systems, the lawyer from Texas offered one of his few substantive contributions to the proceedings. The reason he was silent during most of the meeting was simply that the matters under discussion were so often sufficiently technical that even some of the researchers were occasionally at a loss. During the more heated exchanges, the lawyer's head tended to swivel like a center-court spectator at a tennis match. When he did speak, however, his brief nontechnical contributions tended to be among the more prophetic of the meeting.

"Shouldn't you," he wondered that first morning, "mention something about the control of this work in industry?" This was greeted by a brief silence. Stetten, necessarily, shrugged it off. Such an effort would be "presumptuous," he said—although he certainly intended to distribute the final guidelines to groups like the American Pharmaceutical Manufacturers.

"But we should start to think about this," Stanley Falkow insisted, "*before* people start to do cloning experiments in pharmaceutical companies." Stetten was sympathetic, but could only shake his head. "In that situation," he said firmly, "the Director of NIH really has no more authority than the respect of his position."

That subject was ended. And then an hour's worth of tentative talk on how to certify and rank safe host/vector systems ensued. It was a critical decision: Many of the most interesting experiments in recombinant DNA would almost certainly have to wait at least until a suitable EK2 system had been approved; and if the standards for that system were set excessively high, it could be tantamount to an extension of the moratorium. On the other hand, to set standards too low would be to compromise the whole process in the first place.

And even once the standards were set, precisely who

would be in charge of reviewing the associated survival data and approving each new host/vector system as it came along?

Sydney Brenner, sitting in one corner of the room, warned again that many different kinds of vectors would probably soon be required, and that judging their safety could turn into a full-time job. The point carried more weight here than it had at Torrey Pines: Most members of the committee already felt taxed by their duties, and to sit on a panel issuing safe-vector license plates sounded burdensome indeed. The self-regulation business was clearly not something with a convenient stopping place: Who else, in the first place, could do it?

Stanley Falkow, who would soon follow Berg and Singer in leaving the committee, was the last to speak before lunch. He referred to some recent criticism by pioneer molecular geneticist and gadfly Erwin Chargaff, who in an excoriating essay about recombinant DNA referred to the "Council of Asilomar" and its attendant "molecular bishops and church fathers from all over the world."

"Maybe," said Falkow, "if we start certifying vectors, that's giving too much blessing. Maybe Chargaff was *right* about the bishops of Asilomar."

Straggling back reluctantly from the brief lunch break, no one looked much like a bishop. "We're not moving," Stetten said sternly, "as quickly as we should." In fact, some positions had apparently changed over lunch, solidifying the committee into two voting blocs that would remain permanent throughout the rest of the sessions.

One side was later characterized as "liberal," a potentially misleading designation, since their essential concern was keeping the guidelines as loose as possible. This group was represented most vocally by David Hogness, from Stanford, and Charles Thomas, from Harvard, both respected senior researchers who had performed some of the more advanced work thus far with recombinant DNA technology. In the months to follow, they would become the most consistent focus of the conflict-of-interest charges that the committee would encounter.

Most conspicuous on the "conservative" side were Roy Curtiss, Stanley Falkow, and Wallace Rowe. Curtiss and Falkow, specialists in the genetics of *E. coli*, had both been among the most persuasive voices of caution at Asilomar. Rowe, a youngish-looking man in his late forties, was a more recent committee appointment and as chief of the Laboratory of Viral Disease at NIH had already dealt with some fairly esoteric biohazard questions. Rowe, engaging and open-minded, probably represented the most conservative viewpoint on the committee.

The remainder of the committee tended to adhere to one school or the other. The liberals could usually count on Jane Setlow, the conservatives on Betty Kutter. There remained some swing votes, however, and the results on the more controversial issues tended to be close. If there was, as some critics later suggested, blatant conflict of interest at La Jolla, the conspiracy was well-disguised. The most apparent conflict, without a doubt, came directly from within.

Not too long after lunch, the discussion plunged into the matter of classifying primate DNA that would make me recall that morning's blithe television program about evolution.

How much more dangerous might primate DNA be, compared to the general run of mammals? Since no one knew just what viruses might lurk within the genomes of the lower mammals, the additional hazards of primates seemed unclear, and thus so did the resulting argument. Rowe declared that he would be more comfortable if the DNA came from human embryos, which would be untainted by external viruses. Curtiss composed a compromise that classed DNA from lower primate and human embryos together with the lower mammals; a higher category included postnatal humans and primates.

Sydney Brenner raised his hand. "Where do you put sperm?" he wondered.

"That, sir," Stetten said, "is a very personal question."

But it turned out that Brenner's question was dead

serious. Someone else asked about the DNA from birds.
Betty Kutter suggested raising the containment requirements for all lower vertebrates. In the midst of a series
of voices, Falkow could be heard to say that he was totally lost.

"We keep thinking about harm to man, and forget that
there are many more organisms in the biosphere," Curtiss interjected.

Hogness shook his head at this remonstrance, which
Curtiss had repeated several times before. "We can't
talk about it in such general terms," Hogness said. "Give
specifics."

"We need," Falkow said, "to put some rationale behind these guidelines. That may have been part of the
reason for the uproar already. Education is half the job."

And so it went. The skirmishes on Thursday ranged
from the definition of "pathogenicity" (a narrowly defeated amendment had attempted to change that arguably anthropocentric phrase to "potential ecological
alteration") to a lengthy argument over how to specify
the "purification and characterization of DNA."

The latter issue, in fact, caused a major battle. The
question at hand was, if shotgunning experiments are
actually dangerous, because of the possibility of cloning
undesirable genes, then at what point can one be confident that one has sufficiently purified and identified a
gene sequence to make it safe? The matter was important because some researchers were already talking
about doing their initial shotgunning work in the more
costly high-containment facilities and then taking the
purified gene products, identified as harmless, home to
their own laboratories for in-depth study under lower
containment conditions.

But how does one define the purity of a submicroscopic mixture of gene fragments in the first place? The
discussion started with whether the wording of the
guidelines should put "rigorously" or "highly" in front
of the word "purified" and ended twenty minutes later
in considerable confusion. At about that point, Stetten
interrupted brusquely. "May I inquire," he said, "as to

what we are talking about?" Obscure numbers were suddenly being bandied about. Several researchers had testified that they thought high-level purification, on the order of tenths of a percent, was possible; others had insisted that the notion, at the present level of technology, was worse than meaningless. Paul Berg—apparently an inveterate volunteer—finally offered to write a passage to define how researchers could demonstrate that they had purified and characterized their cloned fragments sufficiently to lower the containment requirement.

Later, however, over coffee in the La Valencia dining room, a representative from the National Science Foundation—which expected to adopt the NIH guidelines also—seemed dubious about such a "demonstration." The NIH committee, he said, seemed to feel that "data is data. And that's true, in terms of the caliber of *this* committee, but in terms of the six thousand grant proposals I read every year, very often data is not data." When, someone at the table wondered, is data not data? "Well," said the NSF official, "if it comes, say, from Ron Davis at Stanford, you'll probably believe it. But what if it comes from Joe Blow in Oshkosh?"

It echoed one of Sydney Brenner's concerns: the notion of proliferation. Brenner had regularly pointed out that there was a double-edged hazard in reducing containment requirements—not only the reduction itself, but also the fact that each step downward makes the work accessible to an increasing number of less sophisticated laboratories. It was a touchy point, and possibly a quick route back to the cries of "intellectual lockout" that had followed the announcement of the moratorium. If a handful of leading researchers sitting at La Jolla recommended containment facilities that Oshkosh U. could never afford—but altogether accessible for Stanford or Harvard—then Professor Blow might indeed feel shut out.

Yet, as Brenner suggested again over coffee, while carelessness can occur anywhere, the smaller and less sophisticated the institution, the more likely the chance for an accident—and also, someone else added, the less chance for a significant amount of peer pressure and scrutiny. Berg wasn't so sure. "Secrecy in doing this work will be impossible in the lab. The student next door will know you're supposed to be using P3, and when he looks in and sees you working on an open bench, he'll be upset, and he'll do something about it." There was a brief silence over the coffee cups, and then a younger researcher pointed out that that's Stanford, where students *understand* what's going on. Berg shrugged, not convinced. Brenner nodded. "I've worked in labs," he said amiably, "where no one in the whole *group* understood what was going on."

The remainder of Thursday afternoon involved complex discussions about specific containment recommendations, culminating in the question of whether the Woods Hole requirements for the lower vertebrates were strict enough. "We're not being conservative enough about the birds and the frogs," one member said, and the table immediately lurched off into a lengthy and occasionally contradictory discussion of just what endogenous viruses might inhabit the genomes of those creatures.

After half an hour, chairman Stetten broke in. "This is a discussion," he said, "that my friend Aristophanes would have liked. He wrote one play called 'The Birds,' and another called 'The Frogs.'" There were polite smiles around the table. "He wrote another play, too," Stetten continued, dead-pan, "and that one was about clouds, and that's where we've been since this morning."

The descent from the clouds followed immediately, however, and involved what would soon become the touchiest issue at La Jolla: the cold-blooded vertebrates. While no one really wanted to argue at length about birds—which were, in fact, ultimately awarded their own category—the realm of the cold-blooded vertebrates was almost certainly the next to be colonized with the recombinant DNA shotgun. Work had, in fact, already begun on the African frog *Xenopus*. And while the Woods Hole version of the guidelines allowed such work with an EK1 vector, the question raised at La Jolla was whether such work might not be safer with an EK2 system—which,

181

judging from what was said at Torrey Pines, would soon be available.

"I've heard nothing but disdain for biological containment from around this table," said Stanley Falkow, a bit heatedly, "but it seems only sensible that we make use of the safe vectors that people have worked so hard to create."

Berg agreed that EK2 should be required for the cold-blooded vertebrates, even though some work had already been done with EK1. "It's a brand-new field, and we should head it off at the pass."

Hogness and Thomas both disagreed. Hogness felt the dangers of shotgunning were overrated; Thomas suggested that the new data on the survival of *E. coli* K-12 in humans was making EK1 look "safer and safer."

The discussion grew heated on both sides, but when the vote was finally taken, the committee agreed, 9 to 4, to keep frog, snakes, and their kin at P2 plus EK1.

After the vote, Rowe broke the tension with a joke: "Well," he said, shaking his head, "I certainly hope I'm not bitten by an *E. coli* with a venom gene." Hogness laughed. "That doesn't seem too likely." Rowe frowned, considered this, and then said, "True. Maybe you'd have to give it the fang gene too."

When the committee reconvened early Friday morning, the same gray fog hung over the coast. The meeting started with some fairly plodding business left over from the day before, but in the middle of the morning, Paul Berg presented his criteria for "purified and characterized" gene fragments. Although they were quite rigorous —and in fact were later adopted without change—Berg seemed less than satisfied, apparently still feeling that the criteria left too much up to the discretion of the individual researcher. Yet, on the other hand, who else was going to do it?

It was the familiar conflict-of-interest dilemma again, perhaps illustrated most clearly that same morning when Roy Curtiss offered his elaborate hand-written suggestions for how EK2 and EK3 vectors should be evaluated.

The broad definition had been that an EK2 system had to pass a broad range of survival tests in the lab, and that EK3 also required testing in actual animal subjects, including primates. Curtiss's recommendations for the EK requirements were extensive, intricate, and cautious —no doubt reflecting his own frustrating efforts over the past several months to disarm *E. coli.* The dilemma arose when Curtiss's modified *E. coli* K-12 months later became the first host organism certified for EK2 status. And thus Curtiss's involvement in drafting thost standards in the first place might be seen by some as a conflict of interest. Yet, at the same time, Curtiss offered a unique and practical insight into the difficulties of disarming host/vector systems, and—in what may become something of a frontier technology paradigm—had he disqualified himself from the proceedings, it would have been difficult to find quite so experienced a replacement.

Not that the committee at La Jolla lacked for outside techincal opinions—but some of them only tended to remind just how far out on the scientific limb this business was. One adviser, a senior Environmental Protection Agency official, regularly and unabashedly offered thumbnail dissertations on basic public health questions. In front of the Torrey Pines group he had elicited audible chuckles with a lengthy description of the outflow from sugar beet processing plants, and out-right laughter when he cautioned that when it came to biohazard control, "you shouldn't really count on your local sewage treatment plant." At La Jolla, the reaction was more restrained when, in the midst of the discussion about endogenous viral sequences in frog and bird DNA, he provided an overview of the bacterial pathogens commonly carried by whitefish.

It's lonely out there on the edge, and also a lot of work. Stanley Cohen—who, while no longer a committee member, was sitting in—spoke from one corner of the room on behalf of "those of us who spend a lot of time sending out plasmids." Cohen's pSC101 had, of course, by then been one of the hotter properties in molecular genetics for nearly eighteen months. It seemed only sensible, he suggested, that if NIH was going to certify host/

vector systems in the first place, they might as well distribute them too. Some fairly costly-sounding solutions were offered, while the NIH officials in the room looked uncomfortable and noncommittal. But then a familiar question arose again: Exactly who was going to certify these vectors in the first place?

There was a brief skirmishing, a vote made it official: For the moment the committee itself would take charge. "No local group," Stetten said supportively, "will have as much expertise as this committee."

Curtiss's EK2 and EK3 descriptions were approved by the committee, and then Stetten wondered out loud whether, considering the difficulties the committee seemed to be encountering, it might not be a good idea to reserve the meeting room for an additional meeting on Saturday. There was a quiet, collective groan, brief mutterings, fervent excuses, and then the discussion trailed off.

It was time, clearly, to have the guidelines finished. And then—with no dissenting opinion—the committee agreed that it was time for lunch.

When the meeting reconvened, attendance had diminished considerably. The gradual exodus of the press had accelerated—the *Time* stringer and the local reporters among the most recent departures—and thus by this time only a handful of science writers remained, most of whom had covered the topic since Asilomar and had come to consider the matter almost their personal property. And it took a personal interest to maintain attention in proceedings that had started, on occasion, to resemble Abbot and Costello's "Who's On First?" routine. Less than twenty minutes after the meeting resumed, one senior NIH official was snoring loudly, head thrown back on a couch surrounded by shelves of conch shells and polished nautilus.

That long final afternoon at La Jolla was a curious blend of ennui and anxiety. The committee members seemed thoroughly exhausted by the subject, and yet, just around the corner lurked the potential for both a calamitous public reaction and some precipitous legislation.

The fear of adverse public reaction never really left the meeting room. At one point, for example, someone suggested adapting the Communicable Disease Center's five-stage classification of pathogenic organisms for use in the recombinant DNA guidelines. "We can't use CDC," someone else pointed out promptly. "They're only interested in natural contagion. CDC Class 2 includes *rabies*. Someone could take that out of context and give all of this a very bad name!"

"The guidelines," an NIH representative pledged brightly, later that afternoon, "will be distributed as widely as possible. Announcements will appear everywhere. And I don't think," he said confidently, "that there will be any problem with receiving petitions for change."

Stetten eyed him coldly across the meeting room. "I certainly didn't have to ask for the fifty-odd letters *I* received after Woods Hole."

Just after that exhange, the committee's quiet Texas lawyer made another distinctly prophetic observation. The issue at hand was local biohazard committees—the groups who would monitor proposed and actual research and who would enforce adherence to the recombinant DNA guidelines. The exact composition of those committees remained a fuzzy issue, but it was clear that most everyone wanted to keep their biohazard committees within their own academic families. "It seems to me," the lawyer said, "that someone from the *outside* should sit on your biohazard committees. You've said that a small institution might have to go outside to get competent people. But I wonder if even Harvard might not be better off with MIT sitting on its committee."

It was only coincidence that it was Charles Thomas —from Harvard—who first objected to the suggestion. "We've been working for two days," he said. "We're near the end, and let's not move capriciously." Thus, the lawyer's notion of an extended biohazard committee was quickly deferred—until, of course, half a year later,

when the City Council of Cambridge, Massachusetts, in-
stituted their own moratorium on recombinant DNA
research and empaneled their own biohazard committee,
which by then included a broader spectrum of views than
simply Harvard and MIT.

For most of the final session, Stetten deftly herded the
discussion through procedural shoals like some odd blend
of gruff schoolmaster and mother duck. And his finesse
worked well until, straight out of left field, the issue of
cold-blooded vertebrates resurfaced.

Early on Friday, the containment requirements for
shotgunning experiments with cold-blooded vertebrates
were raised to P2 and EK2 by a close vote. The widespread attraction of the work, posed against the relatively unknown hazards of endogenous viruses, seemed
to be the factor that had, overnight, changed votes. The
new rating actually represented an upgrading of the
original Asilomar suggestion. The issue seemed settled.
But then, late Friday afternoon, one of the Harvard researchers proposed a "grandfather clause"—a provision
that would allow those who had already cloned cold-
blooded vertebrate DNA in standard *E. coli* K-12 to
keep their original clones, even though they no longer
met the guidelines. There was silence around the table.
It seemed like a dubious idea to draft a clause that allowed some to slip around the guidelines altogether, as if
those who drafted the guidelines didn't think them all
that critical in the first place.

Hogness promptly offered an alternative: There
would be no need for the grandfather clause if the committee simply returned to the previous, lower standard
for the cold-blooded vertebrates. There was a revote,
and the higher rating was sustained. Then it was back to
the grandfather clause. The liberals provided some
fairly eloquent persuasion, and briefly it appeared that
the clause might actually pass. And from the outside, at
least, it looked like a positively suicidal move.

Some voices spoke against the clause. Berg suggested
a phrase that encouraged investigators to put their old
clones in the newer vehicles. Stetten, who usually kept a

low profile, even offered an understated warning: "I forecast," he said quietly, "that this may draw some criticism." Brenner, who had been relatively quiet during the previous two days, was finally very firm about the idea: It was very bad sense to write a blanket revocation into guidelines that are to be read by everyone. He suggested, instead, a sentence specifying that "grandfather" problems could be appealed to the committee on an individual basis.

It was an apt and politic solution. But even so, the problem of self-interest did not remain long submerged. Minutes later, during a discussion of the virus section, Berg suggested tightening a specific requirement. Rowe uncharacteristically demurred, pointing out that the virus field was moving very quickly and that the proposed change would slow down one researcher in particular, who was doing very promising work. It would be a shame, Rowe said, to hold the fellow up for another six months. Berg exploded: That was just the sort of consideration, he said, that would seem to the public self-serving indeed. "You can't tailor guidelines to fit an individual's research." Berg, a virus researcher himself, paused and then grinned. "I'd like to slow him down, in fact." There was laughter, and then from the side of the room Sydney Brenner called out, "Now *that's* self-serving!"

Much later that same night, the La Jolla meeting ended in an almost convivial atmosphere. Outside the conference room door, in the hotel lobby, there was a black-tie champagne party, and the strains of cocktail piano music punctuated those final hours. And just about when the piano player finished "As Time Goes By" and retired for the evening, the last, formal motion to accept the guidelines finally passed unanimously.

Probably the most interesting moment during those evening hours occurred nearly at the end, when Sydney Brenner stood to speak. "I think," he said, "that someone should *do* a dangerous experiment. And I think one should do it in such a way as to find out whether a microorganism *can* transfer DNA into a higher organism. If this can happen, then the upper limit of our caution will likely be infinite." Such an experiment, performed, of course, with P4 containment and the NIH imprimatur, might proceed as follows, he explained: One could graft the DNA from polyoma virus (which causes tumors in mice but does not infect humans) onto some plasmids, put the plasmids into the bacteria, and then feed those modified bacteria to a population of newborn, germ-free mice. One would then monitor the mouse blood for the appearance of antibodies to the polyoma virus, and watch for actual tumor formation as well. A control population of similar newborn mice would be fed the same strain of bacteria without the polyoma/plasmid hybrid. If symptoms of polyoma infection appeared in the test mice, the implication would be ominous: DNA fragments might well be capable of delivering their genetic orders so flexibly that they can still function when altogether removed from their natural vehicle. Such a result, almost certainly, would demand further tightening of the recombinant DNA guidelines. "But regardless of the result," Brenner said, "we will have *numbers*, and that is very important. Otherwise, we have a socially undesirable alternative: retrospective epidemiology in our own laboratories."

But what if the results were negative—if no evidence of polyoma appeared in the subject mice?

Wallace Rowe, the NIH chief virus researcher who would most likely be first choice to perform the experiment, immediately suggested that there would be a real danger of "over-interpreting" a negative result. David Hogness promptly added credence to Rowe's concern: "A negative result," he said, "would be very important, and might make the guidelines almost irrelevant." Rowe leaned back in his chair, arms folded across his chest. "Then I won't do the experiment," he said firmly.

A lengthy discussion ensued, covering every aspect of the issue from how the bacterial infusion should be administered—"Stomach tube, no," Rowe offered, "subcutaneous, in tandem, yes"—to the experimental rationale itself. "It's a very expensive experiment," Berg pointed out, "if we're not all convinced that we'll learn something." Maxine Singer said that she was bothered by how specific the experiment seemed. David Hogness insisted that some kind of experiment was, nonetheless, critical. And there was general concern among the conservatives that if this single "dangerous" experiment proved innocuous, it might provide a justification for indiscriminate flaunting of the guidelines.

In the end, however, it became clear that the experiment was unavoidable. Curtiss moved that NIH proceed "with all due haste," the motion passed unanimously, and Wallace Rowe was put in charge. Brenner said finally, "Let's do this: We'll all put our predictions in a sealed envelope and compare at the next meeting."

13

Recombinant DNA Meets the Public

Long before Wallace Rowe even started to order his germ-free mice, however, a rush of events overtook the matter of recombinant DNA research, culminating in a new wave of public criticism which would quickly garner more publicity and grass-roots support than the NIH efforts had attracted over the whole of the previous two years. It started quietly. But one might well mark February 9, 1976, as the day the opposition began to coalesce—on the occasion of a meeting that, perhaps inadvertently, exerted as much influence on the future of recombinant DNA work in this country as did Asilomar.

The setting was distinctly less romantic than California's Monterey peninsula: It was a chill, gray Monday morning, early in February 1976, and the conference room in which the meeting took place was situated high in one of the concrete monoliths that dot the sprawling grounds of the National Institutes of Health, twenty minutes northwest of Washington, D.C. The official title of that two-day gathering was "Special Meeting of the Advisory Committee to the Director, NIH." What that meant was that for the first time the proposed recombinant DNA guidelines, completed just two months earlier in La Jolla, were finally about to go public. And while the meeting itself, in procedural terms, had relatively little effect on the final form of the guidelines, it seemed to mark the real beginning of organized public resistance to recombinant DNA research.

Not, of course, that the public had been actively excluded during the year of committee deliberations that followed Asilomar. All of the NIH Recombinant DNA Molecules Advisory Committee meetings were, for all intents and purposes, open. But the NIH Advisory Committee meeting was really the first time that public participation was actively solicited. An extensive list of public interest groups, ranging from Friends of the Earth and the League of Women Voters to the National Consumers League and the Sierra Club had been invited; many sent representatives and, in some cases, offered statements. One afternoon of the hearing was devoted entirely to statements from the floor.

The announced reason for the meeting was simple: Now that the Recombinant DNA Molecules Advisory Committee had finally presented their version of the guidelines to NIH director Donald Frederickson, he wanted some public advice on the document—which, for all practical purposes, meant advice from the Advisory Committee to the Director. At that point the director's twenty public advisors included, among others, a chief judge in the U.S. Court of Appeals, the provost of MIT, the chairman of a west coast medical school, the president of the National Consumers League, the president of the National Academy of Sciences, and the chairman of the biology department at Caltech. Compared to the nearly exclusively academic backgrounds of the guidelines committee, this was really almost the public.

That Monday morning it became apparent exactly how interested the public was in this business. Well before its 9:00 A.M. opening, the meeting rapidly filled the largest conference room in the building and spilled over, via closed circuit television, to another audience in a neighboring room. Dominating the center of the main room was an oval conference table thirty-five feet long and half as wide, around which sat the twenty members of the Advisory Committee. Ringing the big table in a horseshoe shape was an audience five deep and numbering at least one hundred, including far and away the largest number of reporters for any recombinant DNA event to date.

Director Frederickson who had, during his seven months in that position, already established a reputation for encouraging public participation—started the meeting with the observation that, from the moratorium onward, the process has thus far "reflected the intent of science to be an open community." Introducing public opinion into that community, he continued, is not quite so simple, but then neither is it impossible.

One would hope not. And the Advisory Committee certainly seemed an apt test case, representing backgrounds that ranged from one gentleman with twenty-five years in medical microbiology to a woman who would soon fetchingly admit that she wasn't exactly sure where "the ballpark" was. The age span covered a similar range, from a silver-haired and dignified old man to one young Harvard student who looked barely out of high school. And while the group may have represented rather more the cream of the public than its cross-section, in one way at least they showed themselves to be characteristic of the public at large when faced with these new questions of molecular genetics: They were bewildered.

Following Frederickson's optimistic introduction, Paul Berg and Maxine Singer proceeded, in quick succession, to dump two massive loads of background onto the unsuspecting committee members. The members had already received a whole packet of background material, mostly in the form of magazine articles, but between Joshua Lederberg's cheerful look in the future in *Prism* and Stanley Cohen's dense *Scientific American* piece on plasmid engineering there had been little that really prepared them for this blitz of data.

Berg's assignment was to sketch the background of the new technology and its potential hazards; at one point, he even used long strings of colored plastic pop beads to illustrate the action of restriction enzymes. Even so, midway through Berg's presentation, some of the eyes of the committee had already taken on a glaze, and most of the new reporters were either scribbling desperately or else writing nothing at all.

Maxine Singer's task was, if anything, even more difficult: She was supposed to outline and explain the proposed guidelines. In the course of an hour-long presentation, she also covered such basic material as the difference between prokaryotes and eukaryotes, and the meaning of 10^9. The amount of territory between those concepts and the more esoteric guidelines was, needless to say, vast.

Hogness and Curtiss then presented their views on the guidelines. Hogness contended that they were too strict and that an "overshoot" past the intensions of Asilomar had occurred. "This was not based on more data," Hogness said. "If anything, data have shown a decreased potential for danger." And Curtiss followed by explaining that the committee had decided that if they erred, they would err on the conservative side. Curtiss said he was pleased with the results of La Jolla, and hopeful for the future of disabled vectors.

By then, the Advisory Committee had heard two hours of dense technical detail, followed by two contradictory opinions from experts who had both voted for the same set of guidelines. After all this, it seemed only fitting that the first subject the Advisory Commitee seized on for discussion was, as the federal judge put it, "the possibility of a major disaster at the high school level."

Berg had inadvertently triggered the issue, in the course of answering a question about the level of competence required to perform these experiments. Berg had carefully distinguished between simply fragmenting DNA with restriction enzymes and the far more delicate matter of inserting those fragments into a host/vector system. Someone asked if the former process could be done in a high school lab. "I suppose so," Berg said cautiously.

And the discussion was off, ignoring the distinction

that Berg had made and pondering the potential threat of high school recombinant DNA. Several of the researchers tried their best to explain. "Even if they *could* do it," Berg said at one point, "they probably couldn't even tell if it had worked or not." The discussion lasted until lunch, however, and was laid to rest only when one of the Advisory Committee members, a microbiologist himself, pointed out that the farthest his high school son had progressed in experimental biology was making cheese from goat's milk, a procedure hazardous, he said, only to olfaction.

It began to appear that the major issues had not exactly been grasped by the Advisory Committee. During the subsequent lunch break, I suggested to DeWitt Stetten that perhaps a two-hour biology lesson wasn't quite enough to prepare some of the Advisory Committee for its deliberations. He snorted and shook his head. "It should have started in grade school."

After lunch the level of discussion rose rapidly from its high school detour. Yet even by the end of the two-day session, it was still difficult to ascertain just what the Advisory Committee was supposed to do; after reading a handful of magazine articles and listening to eight hours of contradictory testimony, the individual members seemed hardly in a position to do anything but nod gravely at the intricate, Byzantine-sounding guidelines that had, after all, taken the recognized leaders in the field a year to assemble.

"Ass-covering," one cynical science writer pronounced the whole procedure. Already there had been rumblings in print about public exclusion from the guidelines process, and some observers felt that the elaborate publicity given to the Advisory Committee meeting was an attempt to make the whole thing look at least a bit more participatory. "A rubber stamp," as another critic characterized the committee.

The characterization wasn't altogether fair. In the end, some of the committee members offered some fairly thoughtful recommendations, particularly in the areas of enforcement and implementation, and some of that thinking influenced the form of the final guidelines. But one could hardly expect the president of the National Consumers League—or, for that matter, the fellow with twenty-five years in medical microbiology—to offer a specific evaluation of, say, the containment requirements for shotgunning experiments with the cold-blooded vertebrates. The real contribution of the Advisory Committee meeting was that it offered for the first time a forum for the opponents of the guidelines. And once the floor was opened for public comment on Monday afternoon, many who spoke expressed their opposition in no uncertain terms.

Nine witnesses had submitted written statements and most had requested time to read them aloud. David Baltimore—one of the original signers of the moratorium letter, and by then a brand-new Nobel laureate—was the first to speak. Baltimore made a strong, eloquent case for the guidelines and the potential importance of the new technology. "In technical issues," he said in reference to the self-interest charges, "it is critical to have public advocates who are specialists."

The second witness was a young graduate student from the University of Massachusetts, speaking on behalf of the Boston Area Recombinant DNA Group—a loose coalition of young scientists, primarily from MIT and Harvard, that had evolved in the months following the Cold Spring Harbor petition and which at that point remained the most active and vocal source of criticism. The young man, the first of the group's several speakers, quickly outlined the three main points they felt must be corrected before the guidelines could be adopted: *E. coli* must be replaced altogether by another host known not to colonize human beings; the containment requirements for many of the shotgunning experiments—especially the invertebrates, like fruit flies—must be raised; and a broad-based enforcement agency must be established to oversee all recombinant DNA work. The speaker was bright, forceful and slightly snide, in the manner that many college radicals in the 1960s would adopt when speaking to an older, middle-class audience. He finished on a loud, emotional pitch: "Even if it takes ten, twenty, fifty-five, even a hundred and five years, I implore you, take the time to tighten these guidelines now."

His style—strident and a bit badgering—was distinctly not right for the sedate Advisory Committee atmosphere. The rest of the opposition speakers, five in all, chose a lower-key approach. Not all the opponents, however, shared the same viewpoint; two, in fact, argued that the guidelines were *too* strict. One of them, a young researcher from North Carolina, presented an elaborate and slightly disjunct case for scientific freedom that wound up quoting Thomas Jefferson on liberty and referring obscurely to the lessons learned from the Spanish Inquisition. The other was Don Brown, the Baltimore researcher who had been the first in the field to work with DNA from the frog *Xenopus*. Brown had in fact provided the purified frog genes that Stanley Cohen and his associates had successfully incorporated into bacterial plasmids in the earliest days of the new technology, and, a year or so after that, Brown had headed up the Asilomar working group on eukaryotic genes. But Brown's future work had been rendered considerably more difficult by the La Jolla ruling on cold-blooded vertebrate DNA; since no EK2 vectors even existed at that point, Brown's work with *Xenopus* would be effectively shut down.

Brown's statement before the Advisory Committee was intense, angry, and cogent. " 'Guidelines' is a terrible word for these," he said. "These aren't guidelines—they're rules that are massive, detailed, forbidding, and above all rigid." He pinpointed accurately the areas in which the guidelines committee had been particularly conservative: It was possible, he pointed out, to work with some whole, infectious viruses under far lower containment conditions than those prescribed by the guidelines for viral DNA removed from its protective vehicle. "Why were such strict and irrational guidelines drawn up?" he asked. "They were produced by scientists who had no choice." And here, strikingly, Brown agreed completely with the critics on the other side of the fence: The problem was insufficient public input, which, in Brown's view, made the scientists bend over backward so as not to appear self-serving. The result was "this kind of rigid, immensely detailed set of rules which encourages people to look for loopholes."

The predominant criticism at the Advisory Committee meeting was, however, just the opposite. A young woman historian from the University of Michigan—who a few months later helped organize an unsuccessful attempt to keep recombinant DNA work off the Ann Arbor campus—argued that the guidelines were flawed from the outset, since there had never been any discussion of whether the work should ever go forward at all. She went on to suggest that the committee should have included no more than two or three people in the field, and that in general, the discussions had "ignored the infinite capacity of living things to change and adapt." Her vaguely vitalistic approach to the science was characteristic of much of the lay criticism leveled at the guidelines; predictably, it did not arouse much sympathy in the hearts of most biologists.

The remainder of the young critics that day had scientific backgrounds and their objections were all similar, focusing on the use of *E. coli* and the guidelines drawn up for shotgunning. There were variations, however; one speaker, for example, feared that the "dangerous" experiment that Rowe was to do with mice and the polyoma DNA was actually a "setup" designed to produce a negative result and a go-ahead for the whole field —an ironic charge considering Rowe's own initial reluctance about the experiment. Other concerns ranged from the possibility that autoclaving might not fully destroy DNA fragments to the fear of the consequences should a hormone-producing *E. coli* escape some pharmaceutical factory and colonize human intestines. One speaker recommended delaying all the experiments indefinitely. "What's the rush?" he said. "After all, we have several billion years of evolution behind us."

Philip Handler, the President of the National Academy of Sciences and a member of the Advisory Committee, remarked later that he was "troubled by the fact that the nay-sayers are far younger than the aye-sayers. Conservatism used to be the role for the elders of the tribe. . . . I'm not sure what it means."

After the first day of hearings, I walked over to the corner of the conference room, where a small group of the young critics had gathered to trade notes. "The mood really changed by the end of the day," said the University of Michigan professor. "I mean, I got the feeling in the morning that everyone was just going to go along, but now they're starting to ask questions!"

Not everyone was so enthusiastic, however. One observer, a bearded man in a ski jacket, told me that the Advisory Committee wasn't "looking ahead past their next meal. They don't know how big this thing really is." He mentioned biological warfare as an example. But aren't laboratory accidents, someone asked, really a far more immediate hazard? "No, no, no," the man replied, a little annoyed. "You've got to deal with it all at once. set up commissions, authorities, regulations. This is a big, big thing." Someone else, tall and thin with an orange daypack, mentioned that he'd already written a book about recombinant DNA but was afraid to publish it because it dealt with possible terrorist applications: "It's so accessible," he said. Later he also made veiled references to organized crime.

My initial reaction that was probably much the same as that of many researchers when first confronted by their young opponents: Are these people, I wondered, serious? They were, in fact, quite serious—so serious that grass-roots opposition to recombinant DNA work, fueled in large part by these very people, turned into a full-fledged movement. And there was real sincerity and feeling behind that movement. But my first impression, at the NIH Advisory Committee meeting, was a sense of déjà vu, a feeling that the campus politics of the 1960s had been plucked out of time and place and suddenly appeared, altogether inappropriately, right here and now. Someone ran up, eyes shining, having just discovered that one of the people in the audience who criticized the guidelines for lower invertebrates was actually a former employee of the National Academy of Sciences. Someone else announced that he'd heard of a paper that showed that naked DNA could survive within a mouse for up to 48 hours. Someone else, inspired, sat down at a table to record all the big business connections of the major names in the guidelines proceedings.

"Such chutzpah," murmured one science writer, referring to the young graduate student who had blasted the Advisory Committee with a series of horror scenarios that the guidelines committee had already been considering for more than a year. She approached him to ask for a clarification of one of his assertions. He bristled: "Are you calling me stupid or something?"

No one was calling anyone stupid—but there was an undeniable, and probably inevitable, information gap between the proponents of recombinant DNA work and the critics. The critics tended to make technical errors and incorrect assumptions—a failing which in these circles seemed tantamount to offering the Modern Language Association a resolution with faulty grammar. One of the more well-informed of the opponents, for example, told the Advisory Committee that all previously made cloned gene fragments should be destroyed, and that "innumerable labs" possessed such clones. This simply was not true, and everyone who followed the field knew it. Perhaps some remnant of the protest style of the 1960s prompted the exaggeration, but, if so, it was an unfortunate influence: This seemed to be one place where hyperbole did not help to make a point.

Besides the information gap, there was one other striking quality to the arguments of the young opponents: The reflex imputation of base motives to the original guidelines writers. The situation was often interpreted as some cabalisic conspiracy to win prizes and create industrial empires at the cost of safety to society. At one point after the first day of the Advisory Committee hearings, one of the Cambridge graduate students buttonholed Wallace Rowe to say that he would have liked to hear Paul Berg and Maxine Singer stand up and *say* what their motivations were in this moratorium business. "I mean," he said, "are they really *sincere* in doing this? I don't know; I'd just like to hear them get up and *say* it." Rowe, with obvious restraint, just shook his head. In the months following, it must have been frustrating indeed for the original instigators of the moratorium to hear their motives so often attacked. In some cases, the earliest participants simply dropped out of the controversy, no doubt feeling distinctly burned. Others continued to take active roles in public debates from Berkeley and San Diego to Ann Arbor and Cambridge.

It was in Cambridge, in fact, at a technical symposium on recombinant DNA just prior to the issuance of the final guidelines, that I finally heard what seemed the archetypal gut reaction to the situation. It was just a few hundred feet from the conference auditorium in an MIT meeting room, where the group called Science for the People had called a lunchtime meeting on the guidelines. The issue that had brought them together that day was how, specifically, to stop them. The meeting room was packed with at least one hundred people, but the only old hand I could see from the recombinant DNA committee was Stanley Falkow, sitting rather sardined into one corner. The discussion took a familiar course: a harsh analysis of the NIH guidelines and the motives of the researchers involved, followed by a bizarre series of charges and contentions from the audience. The latter appears to be an inevitable companion to the hard-line populist approach to science; one is, apparently, obliged to entertain cheerfully almost any amount of pseudo-scientific blather from the floor. Finally, however, after listening to nearly an hour of nonstop excoriation of everyone involved in the guidelines process, Falkow—usually rather a soft-spoken sort—apparently could take it no longer. He stood up from his seat in the corner and asked loudly, almost sadly: "What I'd like to know is, where the hell were all you people when this thing *started*?"

Good question. And one with less than a simple answer.

A few months after the Advisory Committee meeting I, visited a young postdoctoral fellow named Richard Goldstein in his small office at Harvard Medical School. Goldstein had been one of the originators of the Cold Spring Harbor petition, a member of the Boston Area Recombinant DNA Group, and probably the most effective and articulate of the critics who appeared before the February hearings of the NIH Advisory Committee.

Goldstein had returned from those hearings hurt and a bit angry. The Boston Area Group had, after all, spent two months composing a long and carefully detailed critique of the Woods Hole guidelines, which was sent to each of the members of the NIH Recombinant DNA Molecules Advisory Committee prior to La Jolla. "I never heard a word from any person on that committee other than Stetten, who said thank you. I don't have the faintest idea whether it was even considered." Goldstein shook his head. "And even after the Advisory Committee meeting, I *still* don't know."

As one of the most well-informed of the critics, Goldstein was occasionally embarrassed by some of his unsolicited allies, as at the Advisory Committee hearings. "I wish," he said, "they'd stick to the issues they know." Yet Goldstein also felt altogether cut off from the guidelines process itself. "You can close it off," he said, "if you hold your meetings in places like La Jolla, which is probably the furthest corner of the country from where all the opposition is." He shook his head. "I always try to put myself in the other's position, but after these things keep happening over and over again, you get to feel sort of paranoid." His interpretation of why the meeting was held in La Jolla probably was a touch of paranoia; when that meeting had been planned, in fact, Roy Curtiss had offered to help host it in Alabama, to which the response was *"Alabama?"* While the choice of a beautiful beachfront hotel in California was doubtless not entirely random, neither was it particularly Machiavellian. Some of Goldstein's other criticisms seemed closer to the mark.

The selection of magazine articles given as background material to the members of the Advisory Committee was, without question, favorably biased and did not represent the full range of opinions that had already appeared, even in journals as restrained as *Science*. Goldstein felt also that the structure of the meeting—with proponents like Berg and Singer speaking at length in the morning, and the opponents left with only short amounts of time late on a long day—was unfair, and that Frederickson had been unnecessarily condescending in introducing the young opponents. David Baltimore, on the other hand, had been conspicuously introduced as a new Nobel laureate (to which Baltimore, to his credit, responded quietly, "I don't think that's relevant here").

In scientific circles, of course, such credentials are in fact relevant, and if one's rank is high enough one can rise above the usual red tape that comes with involvement in public processes. After the Advisory Committee meeting, Goldstein was uncertain whether he could continue his own involvement. The long critique had taken much time and now the heads of his department were pointedly suggesting that he make a choice between scientific politics and his own work.

After we'd talked for an hour or so that afternoon, the abrasive, contentious approach that Goldstein had adopted in public had vanished. In odd ways, in fact, he'd come to remind me a bit of Paul Berg. "I'll be sad if the guidelines don't work out," he said. "I think you can work within the system; I don't know if that's an elitist position, but then I have no great faith in government control over anything. That's why I felt that somehow we had to regulate ourselves."

Yet by that time Goldstein clearly felt himself to be irrevocably on the outside. "It's one of the strange strains throughout this," he said finally. "There were plenty of people who knew we were deeply involved in this, and they had a lot more at stake than we did, because they wanted to do the work. Yet they never picked up the phone and said, gee, let's talk about this thing. It seems like that would be the quickest way to get it straightened out—so they'd know where they were going to work for the next year or two."

The criticisms raised by Goldstein and company seemed to have had a minimal effect on the Advisory Committee. The second day of comments provided just about the reaction one might have expected from the group: general agreement with the guidelines, a handful of specific suggestions about enforcement, and regular exhortations toward continued caution. "NIH," said NAS president Handler, "is to be congratulated on getting this far. The question is where you go now." What troubled him most, he said, was that he had found it possible to sit through one-and-a-half days of controversy "and agree with almost everything I heard."

"If Dr. Berg and his colleagues don't deserve the Nobel Prize for medicine," said another member, "they deserve it for Peace"—he paused—"since they don't give one in controversy."

"There's no instant feedback here," lamented one scientist. "With toxic chemicals, it used to be easy: You keeled over."

Other comments expressed real alarm, particularly about the effect public opinion might have on legislation. "You can pat yourselves on the back," said the president of one student organization, "but keep looking over your shoulder. If Congress gets the idea that scientists aren't playing straight, you'll be using your pipettes to pick your teeth."

All of the committee members were to submit written opinions, and Frederickson announced that he would make his decision on the basis of those and the hearing transcript. But DeWitt Stetten probably had the last word in terms of summing up the two days' proceedings. He described a fog-bound strait that separates Newfoundland from Canada—a narrow, treacherous passage which sailors navigate by listening for the sound of the surf on the left side and the sound of the surf on the right. When the sound from each side is equal, then they know they are in the middle, and they sail on. "And that," Stetten concluded, "is my best advice to you."

NIH Director Frederickson smiled and shook his head. "Maybe," he said, "that explains the roaring in my ears."

14

The Frankenstein
Syndrome

My generation, or perhaps the one preceding,
has been the first to engage, under the leadership of
the exact sciences, in a destructive colonial warfare
against nature. The future will curse us for it.
　　　　　　　　　　—Erwin Chargaff, letter to
　　　　　　　　　　Science, June 1976

Recombinant DNA is the most overblown thing
since your brother-in-law created the fall-out shel-
ter debacle.

　　　　　　　　　—James Watson, to Sargent
　　　　　　　　　Shriver during New York
　　　　　　　　　State hearings on recombinant
　　　　　　　　　DNA, October 1976

If NIH director Frederickson's ears were already ringing
in February, by the early summer of 1976 he must have
been positively deafened. Between the Advisory Committee meeting and the final, late June issue of the
first thick guidelines booklet, the matter of recombinant
DNA rapidly and noisily achieved the status of full-
fledged public controversy.

The most visible center of that controversy was, as
mentioned earlier, Cambridge, Massachusetts. Even be-

fore the final guidelines were issued, Mayor Alfred
Vellucci called—only too happily, perhaps, in light of
his traditionally adversary relationship with the local
universities—for a two-year moratorium on all recombinant DNA work within the city limits.

The mayor promptly convened a special City Council
meeting to consider the matter. The dispute lasted
through two bitter public debates, each stretching well
past midnight, and after a complex display of three-way
head-butting between the city, opposed scientists, and
recombinant DNA proponents, a compromise was
reached: A three-month moratorium, along with the creation of a body of scientists and citizens to investigate
the entire issue and report back to the City Council.

Little light was shed during the public debates. "Now
I know," said one scientist who participated in the hearings, "what the Scopes trial was like." One council-
woman said, "This is something that science is just
going to have to learn to live with."

The Cambridge action—which at the most popular
level included pro- and anti-recombinant DNA booths
at a local summer street fair—seemed a precedent-
setter. The Ann Arbor, Michigan, City Council had
months earlier been approached with a similar moratorium petition, but in that case the City Council deferred
judgment to the University of Michigan Board of Re-
gents—who, after considerable debate and a two-day
symposium that attracted more than 2200 spectators,
approved the construction of two P3 facilities and recombinant DNA work at that level. It was a brief victory. By the summer of 1976, the Democratic Party of
Washtenaw County, Michigan—which includes Ann Arbor—adopted a resolution that research on DNA recombinants should be limited to a few isolated facilities.
And Mayor Vellucci, fresh from his temporary victory
in Cambridge ("I have learned enough about recombinant DNA molecules in the past few weeks," he told
one interviewer, "to take on all the Nobel prize winners
in the city of Cambridge"), promptly offered a meeting
of the United States Conference of Mayors a resolution

prohibiting any recombinant DNA work in any city until a public hearing had taken place.

And that was only the beginning. By the end of the summer, San Diego, California; New Haven, Connecticut; and Bloomington, Indiana, were all well on their way to similar local proceedings. And in early fall the state's attorney general and his environmental health bureau, on the same topic. By then, the cast of expert testimony had grown distinctly familiar—James Watson and David Baltimore on one side; Erwin Chargaff and Science for the People's Jonathan King on the other.

At the New York hearings, Baltimore suggested that those who mistrust scientists as irresponsible, prize-seeking individuals (perhaps, he hinted, inspired by Watson's candid book *The Double Helix*) are misled. "It is not true of any of the labs I know. I think that the scientific community, by being as open as it is, and as self-critical, provides a better guarantee of safety than does any government regulation."

"We are jumping into the middle," Chargaff responded, "and assuring everyone that it is OK. I think it is not. I think we are very well able to do a lot of damage to ourselves and especially to following generations."

In the months following the Advisory Committee meeting, Chargaff and Robert Sinsheimer, the head of Caltech's biology department, rapidly became the most influential—and senior—critics of recombinant DNA work. Chargaff was already firmly established as the gadfly critic of molecular genetics, consistently more outrageous even than James Watson, his closest competitor.

Tht issue of recombinant DNA brought out Chargaff's most contentious side, even though it had been his own repudiation of the tetranucleotide description of the DNA molecule, back in 1948, that had been an important cornerstone for molecular genetics. But in a long letter to *Science* early in June 1976, Chargaff blasted the guidelines procedure. "I don't think," he wrote, "that a terrorist organization ever asked the Federal Bureau of Investigation to establish guidelines on the proper conduct of bombing experiments." He continued with an attack on both *E. coli* as a host organism, and the constitution of the recombinant DNA molecules committee. "Our time is cursed with the necessity for feeble men, masquerading as experts, to make enormously far-reaching decisions. Is there anything more far-reaching than the creation of new forms of life?"

Chargaff's letter concluded with a succinct statement of the deepest fears of all opponents of recombinant DNA work. "This world is given to us on loan. We come and we go; and after a time we leave earth and air and water to others who come after us." The question, as he saw it, was simple: "Have we the right to counteract, irreversibly, the evolutionary wisdom of millions of years, in order to satisfy the ambition and curiosity of a few scientists?"

Chargaff's unflinching condemnation provided effective fuel for the many pamphlets, speeches, and symposia that were to follow. But the criticism that proved most unsettling within the scientific community itself was that of Robert Sinsheimer, the soft-spoken head of the Caltech biology department. Sinsheimer had been one of the first biologists to see in recombinant DNA the seeds of a far larger human dilemma. A few months after Asilomar, the British journal *New Scientist* published an article by Sinsheimer called "Troubled Dawn for Genetic Engineering," wherein he asked the following questions: "How far will we want to develop genetic engineering? Do we want to assume the basic responsibility for life on this planet—to develop new living forms for our own purposes? Shall we take into our hands our own future evolution?"

The questions themselves were not new. But for the first time they originated not with isolated think-tankers, but from a scientist in the field itself, evaluating work in progress. And Sinsheimer's broad concerns would continue, but by the time of the NIH Advisory Committee hearing, he had focused them onto a specific technical issue—something called the "barrier theorem." According to this theorem, there may well be a fairly crucial reason for the division of all living organisms into the two great groups of the lower prokaryotes and the higher, nucleated eukaryotes. While prokaryotes, like bacteria, regularly wreak havoc with the internal functioning of eukaryotes, like human beings, Sinsheimer argued that there has never been, in the past, any real genetic transfer between the two groups. If, hypothetically, some eukaryotic DNA control sequence was grafted into a prokaryotic parasite like a bacteriophage, might it suddenly prove that such a modified phage, previously only a threat to bacteria, would find itself able to colonize human cells as well? The notion, though purely hypothetical, was very disturbing. Yet if the theory was correct, recombinant DNA research just might provide that sort of irrevocable genetic bridge.

The theorem was quickly taken up by the critics of recombinant DNA as an especially legitimized adjunct to previous disaster scenarios. And there was an immediate flood of counterexamples from the pro-recombinant DNA scientific community, ranging from the suggestion that intestinal bacteria already take up fragments of digested eukaryotic DNA (delivered in the form, say, of a New York strip steak), to the instance of a specific plant tumor which may be induced by plasmid DNA.

While Chargaff's rhetoric had provided good copy for the popular press, Sinsheimer's specific concern—whether justified or not—once again underlined for the scientific community just how uncertain, uncharted, and tentative this business was.

But neither Sinsheimer nor Chargaff, nor any other critics for that matter, found themselves without ready outlets in the national press in the months preceding the publication of the final guidelines. As concern over recombinant DNA grew, so did the number of reporters and publications who covered the story—accompanied by a steady decline in the quality of that coverage.

The decline started with the Cambridge controversy and the introduction of the phrase "creation of life experiments," which, while grossly inaccurate, proved sufficiently infectious that it rapidly turned up in both the *New York Times* and the *Washington Post*. During the same dispute, the *Washington Star* simplified the question even further, with a headline that asked "Is Harvard the Proper Place for Frankenstein Tinkering?" Much of the media garble was simply a function of reporters with a limited scientific background trying to cover an issue that, by then, probably qualified as one of the most baffling technical mazes of the century. Some treatments, however, did not seem so benign. A feature article in *Ms.* magazine, for example, published just when the Cambridge City Council was holding its hearings, blithely confused recombinant DNA techniques with an altogether different process called cell-fusion, dubbed the product a "genetic stew," and suggested that this nonexistent procedure would result in a "doomsday bug." The article, titled "Genetic Engineers—Now That They've Gone Too Far, Can They Stop?" would not be noteworthy as other than an apt example of technical ignorance in the guise of journalism, were it not for the disheartening fact that the author herself was a professor of communications at a large American university.

Little wonder, then, that many of the principals in the recombinant DNA controversy grew increasingly reluctant to return telephone calls from journalists. A handful of more media-oriented researchers, however—particularly those staunchly opposed to recombinant DNA work—suddenly found themselves with a new and attentive audience.

One researcher, for example—a professor with twenty-five fears of experience in molecular biology—wrote a piece on recombinant DNA research for the *New York Times Magazine*. He made his opposition to the work clear at the outset, suggesting, for example, that the Nobel committee should decree that no awards would ever be given in the field. Yet, subsequently, it was difficult not to wonder at the inconsistency of his reporting. The author cited both the exact year and the early discoverers of restriction enzymes (long before their experimental utility was even suspected) but further on, apparently found it unnecessary to distinguish between the laboratory-enfeebled strain of *E. coli* K-12 and the wild-type *E. coli* that commonly colonizes humans. Even the closest lay reader would probably have to conclude that recombinant DNA research had thus far proceeded with the very same strain of bacteria that inhabits the public gut, and that consequently, the scientists involved were irresponsible men indeed.

That wouldn't have been, of course, an altogether fair conclusion. It underscored, however, the power of the "expert" opinion and interpretation, and called to mind one of the suggestions offered by the University of Michigan professor at the Advisory Committee meeting—her notion that the recombinant DNA molecules committee should have included only three or four molecular biologist out of fifteen members.

The notion, in retrospect, seemed naive. That's not to suggest that nonscientists shouldn't sit on such committees; if there is any purely political lesson in the recombinant DNA situation, it is that they should, and from the very outset of any potentially controversial technical question. Yet if such a committee was composed predominantly of laypeople and included only a minority of "experts," it's not difficult to guess who would have the final word on any scientific dispute—and who could, through the selective interpretation of information, exert inordinate influence on the committee's deliberations.

Thus, the need for a great deal of expert testimony and opinion. Although, as the recombinant DNA situation went public, it also aptly reflected Oppenheimer's observation on the early days of nuclear energy: In the new realm of synthetic biology, it seemed difficult for anyone to claim the final word. And so, in the months after the Advisory Committee meeting, recombinant DNA rapidly became a fertile substrate for all manner of public paranoia.

The young man at NIH who had already managed to see the fingers of organized crime in the situation was only a hint of the bizarre predictions that were put forth. My own favorite during the summer of 1976 was one exceedingly anxious gentleman from southern California who, armed with a mailing list, a mimeograph machine and a Ph.D. of unspecified nature, cranked out a series of releases capped by one sheet entitled "Hypothetical Scenarios Illustrative of Recombinant DNA Hazards." His scenarios, worthy of a Kurt Vonnegut, ranged from a Nobel-prize-winning molecular biologist who threatens, in his acceptance speech, to decimate the planet with his own fatal version of the common cold unless the nuclear powers unilaterally disarm, to the head of a central African state who announces, over satellite television, that his country has perfected an "ethnic weapon" germ that causes cancer in 99 percent of all Caucasians. That, as it develops, is the good news; the bad news is that the deadly microbe has managed to escape from the laboratory and is already spreading, via the monkey population, across the continent.

Early that same summer, one important question about the future of the new technology received a first, tentative answer: In June, researchers from Stanford and the University of California at Santa Barbara reported that DNA derived from bakers' yeast—a eukaryotic organism—had been used to successfully program the prokaryote *E. coli*. The experiment had involved a mutated strain of *E. coli* constitutionally unable to make a certain necessary enzyme for itself. Bacteriophage vehicles, carrying the appropriate section of yeast DNA, were sent into the deficient bacteria and, as nearly as anyone could tell, the yeast DNA was incorporated and began to program for the production of the missing enzyme.

It was the first published evidence that eukaryotic DNA could function within prokaryotes, and while the product involved was fairly basic, it lent considerable credence to the more promising of the early pharmaceutical applications for recombinant DNA. And it also, of course, provided another piece of evidence for biohazard scenarios.

One of the more entertaining scientific symposiums of the year occurred early in June also, just two weeks before the final guidelines were released. The occasion was the Tenth Miles International Symposium, a conference

sponsored annually by the pharmaceutical company of
the same name, and customarily centered on whatever
constitutes the hot biomedical topic of the year. In
1976, the hot topic was obvious: "Impact of recombinant molecules on science and society," as the program
title described it. The organizers, clearly, saw the emphasis on science—but the symposium was held at MIT,
precisely in what then constituted the heart of anti-
recombinant DNA territory. It was that same week, in
fact, that a local newsweekly published the Harvard
biohazards story that led soon after to the Cambridge
moratorium.

Both Science for the People and the Boston Area Recombinant DNA Group were much in evidence—the
former with booths outside the auditorium and regularly
scheduled seminars during the Symposium's off-hours.
And Science for the People managed also to garner the
majority of newspaper coverage—particularly when, at
one lunchbreak, their representative refused to vacate the
stage, yielding only when a Miles organizer threatened
to shut off the PA system. "Scientists Wary on Research"
was the way the *Washington Post* headlined the symposium, noting that "instead of pressing forward with new
research, some of the nation's brightest young scientists
are pulling back."

That particular symposium, then, was probably not
the best possible public relations investment for a corporation that, after all, was in the business of manufacturing restriction enzymes itself. But it nonetheless produced
some dramatic moments: as, for example, when close to
the end of the symposium, someone near the back of
the big MIT auditorium stood and asked, in halting
English, if the rumors in Paris were true: Had the
Americans taken out patents on recombinant DNA?
There was a brief silence. The person at the podium
shrugged. "Is there anyone in the audience who knows
something about this?" After another moment, Stanley
Cohen stood and announced that, in fact, the rumors
were true: Stanford and the University of California had
both applied for patents on the basic work published in

1973 and 1974 by Cohen's and Herbert Boyer's labs—the work which had first demonstrated the feasibility of cloning foreign gene fragments within simple bacteria. The patents, Cohen explained quickly, would cover only the commercial use of recombinant DNA techniques within the United States. And they might, he added hopefully, even provide a means for extending the authority of the NIH guidelines: Perhaps commercial users of the process might be required to sign an agreement to abide by the NIH rules.

Probably the most interesting aspect of the symposium, however, was one that occurred only as a matter of convenience, when one evening, two review panels met in small MIT conference rooms to consider the first of the modified hosts and vectors for possible EK2 rating. In one room, a panel of five researchers examined Roy Curtiss's EK2 *E. coli*; in another, a similar number looked at a modified bacteriophage. The group considering Curtiss's *E. coli* included only two members of the guidelines committee; the others were researchers with appropriate backgrounds in bacterial genetics. Curtiss and an NIH official sat at the small conference table as well —Curtiss, sitting slumped back in his chair and chewing occasionally at one thumbnail. He had submitted a document fully an inch and one-fourth thick, packed with charts and diagrams. In addition, he had covered the room's blackboards with a whole set of survival tables that ranged from "Rat Fed 7×10^9 by Stomach Tube" to "Dessication." The audience numbered about thirty, and ranged from an MIT archivist to a few radical opponents.

The meeting had the benign atmosphere of a Boy Scout merit badge hearing, where a group of fathers would sit in solemn silence in some grade school classroom and pass judgment on whether the scout in question had faithfully fulfilled the requirements for the badge at hand. Only, in this case, both the badge and the requirements were brand-new quantities; as much of the discussion concerned how to interpret the guidelines as how to apply them.

The talk ranged from the survival rate of the modified *E. coli* in germ-free mice to one reviewer's query as to whether Curtiss might not actually be able to "beef up" his enfeebled bacterium a bit, so that it could grow more quickly in the lab. The small audience offered a handful of comments as well, from an extensive mimeographed critique by the Boston Area Recombinant DNA Group, who felt most emphatically that Curtiss's efforts did not qualify as EK2, to one young woman, a self-described student of ecology, who was worried that gulls might retrieve the modified *E. coli* from the Charles River. No one, she pointed out firmly, had done survival tests with either Charles River water *or* seagulls.

Curtiss and Stanley Falkow both provided long, reassuring answers to the latter query; this was something, they said, that had been considered ever since Asilomar, and the general thinking was that if a modified bacterium couldn't survive in the ideal environment of the germ-free mouse, it would have a difficult time in, say, the Charles River.

The Boston Area Recombinant DNA Group critique was not quite so amenable to a reassuring answer. The paper, twenty-odd pages long, was intricate, technical, and carefully reasoned. It included, moreover, a covering letter from Caltech's Sinsheimer to Richard Goldstein, the chief author, commending the critique as "thorough and meticulous." For procedural reasons, however, the critique had apparently not reached the review panel members until very recently. When a representative of the Boston Area Group stood to say that the report needed a full reading to be appreciated, the head of the panel shrugged. "Well," he said, "we all read it, either rapidly or slowly, depending on when we got it today."

In the end, the critique seemed to have little effect on the proceedings. At last Curtiss was asked to leave the room for the actual voting. The process seemed a formality—everyone there had already heard much about Curtiss's work in the previous months—but even so, it was not altogether effortless.

"What Roy has done," said Stanley Falkow, "is monumental, staggering. But it opens up the door to certain experiments. It's a serious thing. Some of those experiments are probably already being done with EK1, God knows . . . so I'm torn. But I have reservations."

The rest of the panel tried to coax him. "You have reservations about the whole concept of biologicol containment!" one NIH representative chided. But Falkow would not budge. When the vote "to recommend certification" came up, there were four ayes and one abstention. The NIH official promptly reworded the motion to "meets the criteria of the discussion"—and only then, with reluctance, did Falkow vote yes.

Neither Falkow nor the Boston Area Group need have worried so much. EK2, as it developed, was not around the corner. Just down the hall, in another MIT classroom, the EK2 phage candidate was disapproved and sent back to the drawing board. And while Curtiss's *E. coli* was approved, no accompanying plasmids or phage existed—and so, in the absence of a fully tested host/vector combination, it would still be many months before EK2 work could even begin.

A few hours after the meeting of the safe host and vector working group concluded, I saw Curtiss at a small party in a Cambridge apartment, and told him that his *E. coli* had passed muster. He shrugged. He was worried, he said, that some might think the panel had been a stacked deck. But he was even more concerned that this particular *E. coli*, the product of so many months' research, might someday be considered for EK3 status.

"I hope not," he said. "We can do much, much better now."

One recurrent topic at the Miles Symposium was the international reaction to recombinant DNA research. Already there were stories of European researchers who had packed up their experiments and crossed borders in order to perform work that their own labs would not allow.

The international picture was nothing if not diverse. Just as NIH was preparing to release its guidelines, Australia, had already published its own set of voluntary guidelines, very similar to the American version. Both Canada and West Germany were well along in their own guideline formulations. France, Japan, the Netherlands, and Sweden were in earlier stages of similar considerations.

Many nations were curious to see just how the United States would handle the situation; others, however, displayed little inclination toward drafting guidelines at all. The Soviet Union and some eastern European countries, in particular, showed no interest in guidelines, even though the Soviet Ministry of Health acknowledged that the U.S.S.R. was actively involved in recombinant DNA research, both basic and applied, at a number of scientific institutes. More than one American biologist would later suggest that, with the expensive guidelines in the United States, recombinant DNA work might finally provide the U.S.S.R.'s opportunity to catch up on the years of biology lost during the era of the scientific dogmatist Trofim Lysenko.

In the end, only Great Britain undertook an extended guidelines effort similar to that of NIH. And although its final Report of the Working Party on the Practice of Genetic Manipulation would not be established until the end of August 1976, it would display two major differences from the American approach. For one, the British guidelines—while intended for voluntary observation— could be implemented by that country's Health and Safety Commission, which, under a law much like the United States' Occupational Safety and Health Act, would have power to enforce those guidelines not only in universities, but in government and industrial laboratories as well. The British guidelines, moreover, did not firmly assign given experiments with given organisms to specific categories; while the report included "suggested" categorizations for typical experiments, each proposed piece of work was to be individually rated by a new body called the Genetic Manipulation Advisory Group.

The British guidelines, while they probably generated considerable scientific red tape, avoided two of the major criticisms that the American version had drawn: the lack of control over industry, and an excessive specificity in the face of an altogether unexplored field. But even the British did not manage to escape a third, and potentially most damaging criticism.

While the British process had included testimony from a far broader range of witnesses than had the NIH approach—down to the level of laboratory technicians and glass-washers—there were still charges in the British press of insufficient public participation. *Nature*, the most prestigious of British science journals, even offered this editorial observation: "Thus far it has certainly been possible to state a view and have it considered by those debating the issues—but there has been nothing like the openness offered in the United States."

Nature's praise, in the summer of 1976, must have sounded more than a bit ironic to some of the people at NIH. And the "openness" had really only begun when, on June 23, 1976, NIH finally released the first public fruit of the Asilomar Conference: the final draft of the NIH "Guidelines for Research Involving Recombinant DNA Molecules."

The guidelines, plus associated appendices, ran 181 pages, with another twenty-seven pages devoted to NIH director Donald Frederickson's own analysis of the situation. At a well-attended press conference at the NIH headquarters in Bethesda, Frederickson presented an additional prepared statement, and then, along with a panel including DeWitt Stetten, Emmet Barkeley, and Maxine Singer, fielded questions from the press.

Most of the subsequent questions were not new. A television crew, however, was in evidence—that medium's first appearance since the press conference following Asilomar. "Can I ask you about hazards?" the television newsman asked Frederickson, as he signaled his crew to start lights and camera. "Could you give us an example of a *specific* hazard that might occur? What's the worst," he prompted, "that *could* happen?"

Frederickson dodged the question cautiously. "It's

rather easy to construct imaginary scenarios," he said, "but we have no basis for sensing their probability."

The television reporter was undaunted. "Is there something that could happen," he asked sonorously, "which could be a worldwide health hazard? Could you tell me about that?"

Frederickson, now maneuvering carefully to avoid becoming a ten-second spot, blurting something about cancer or epidemics, on the six o'clock news, provided a lengthy answer that centered on the question of DNA expression in foreign hosts and sounded suitably unamenable to even the most skillful television editing.

The television reporter decided to give it one more try. "What is the greatest fear of the scientific community itself? Can you tell me that?"

Frederickson paused briefly. "I think," he said finally, "that their greatest fear is the tremendous potential power in his technique. They want to use that power very conservatively, with concern not only for themselves, but for the public and the environment. I suppose," he said, "that the greatest source of fear is really the unknown."

15

The Reins of Evolution

In 1947, Hollywood produced an unusual film called *The Beginning or the End*, a thoroughly fictionalized account of the Manhattan Project and the birth of atomic energy. The last reel takes place on an American-held island in the south Pacific, where the first nuclear bomb is to be assembled prior to the attack on Japan.

In the course of assembly, two pieces of fissionable material accidentally come together (an incident vaguely patterned on the more deliberate Los Alamos exercise called "tickling the Dragon's tail"), and the device promptly approaches critical reaction. In a heroic act, one of the fictionalized young scientists reaches barehanded into the glowing heart of the bomb and pulls the masses apart, saving not only the war effort but the island and all its bivouacked GI's as well. He has exposed himself, however, to a fatal dose of radiation.

He lives only another twenty-four hours, but before he dies, he drafts a letter to his pregnant wife, which is carried back to Washington, D.C., by another fictionalized young scientist. The second scientist takes the new widow to the steps of the Lincoln Memorial before he hands over the letter, and thus it is against that imposing backdrop that she reads it—accompanied by the voice-over of her late husband.

The letter is in a noble style, expressing the young scientist's belief that he has not died in vain, and concluding with a stirring paean to the peacetime potential of nuclear energy. Someday, the voice-over whispers from somewhere in the upper recesses of the Lincoln Memorial, his son will drive from coast to coast, in an automobile powered by the atoms in a single blade of grass. . . .

It's difficult to imagine the first popular movie based on the birth of synthetic biology—except, of course, to predict that it will neither involve the narrative drive of a world war nor the trenchant symbolism of critical masses pulled apart with bare hands.

But one thing is clear: Regardless of the regulatory action taken by any city, county, state, or nation, the story has already begun. And the work will show no more respect for human boundaries than has nuclear energy. The potential rewards are too great and the problems of the planet too far advanced to allow us the luxury, at this late date, of turning our backs on any technologic fix.

That may well prove to be our undoing. But in a curious sense, the new biology has finally completed a circle between science and nature. One can no longer shun science in order to grow closer to nature—because suddenly, science bids fair to put us in the position of being nature itself.

Our species stands at the edge of a remarkable evolutionary precipice, from which we could either fall or learn flight. In years to come, human beings will find themselves in charge of evolution on this planet, not only in the traditional role of destroying old species, but also in the altogether novel business of creating new ones.

That prospect recalls one of Robert Sinsheimer's early criticisms. "There was," he said, "at Asilomar, no explicit consideration of the broader social or ethical implications of initiating this line of research—of its role as a possible prelude to longer-range, broader-scale genetic engineering of the flora and fauna of the planet, including, ultimately, man."

Sinsheimer was correct. Yet that single-sightedness

was also just how Asilomar had been intended. The larger questions were set aside at the outset to deal with what seemed the most pressing matter: How, most safely, should the early research proceed?

That question alone was enough to fill four days at Asilomar, and unnumbered days thereafter. But the approach also gave rise, in the months following the conference, to the idea among some scientists that the recombinant DNA issue had attracted undue public attention—that the matter had been a "technical decision" from the outset, solely involving matters of laboratory technique and good scientific judgment.

Considering the vociferous nature of that public attention, the scientists' sentiments were perhaps understandable. But their analysis was not. It missed, in fact, the key lesson of the recombinant DNA controversy— that this is precisely how the future will happen; in terms of tiny, incremental "technical decisions."

The progress of synthetic biology will never again simply involve pure science. Each decision, each new technique, each step forward will carry its own rider of ethics and responsibility; and the sum of those tiny increments may someday represent one of the most profound ethical and practical judgments in the history of the species.

Considering the reaction to recombinant DNA, chances seem good that the public will likely never again be satisfied to hear that a matter under discussion is strictly technical. One can only imagine, for example, the controversy that may ensue when the first experiment is proposed that uses a virus vector, capable of infecting human cells, to carry a specific segment of human DNA into a patient genetically deficient in a given biochemical function.

Such an experiment is no longer altogether unforeseeable. And its implications are legion. By the time those kinds of questions arise, society will require a new sense of the public as participant. The necessity for early and full public participation was clear in the aftermath of Asilomar—a meeting which, in some history text of the distant future, may even be interpreted as an inadvertent call for help.

No longer can the public representative in scientific decision-making take the part of foreign emissary. There should be little excuse for technical illiteracy in the years to come; science can no longer be an optional duty, or a hobby, or a magazine subscription. Each step into the future should represent a species-wide decision—because, in the end, everyone has a genome.

But some ideal form of public participation is only half the answer. The other half involves changing attitudes among scientists. No longer can the scientist view public education and the occasional ordeal of the newspaper interview as something simply to be endured. The scientist has already been held accountable after the fact; now he needs to be seen as responsible beforehand.

The alternatives are bleak. I still recall the bitter assessment of one young molecular biologist who suggested in a black moment that by the turn of the century either the public will rise up and burn the laboratories, or else money will be shut off, young people will avoid science from the outset, and the discipline will suffocate.

The young biologist's disaster scenario provides a curious counterpoint to the recombinant DNA disaster scenarios that circulated in the months following Asilomar. Come to think of it, with so many scenarios already in existence, someday it's all bound to make a remarkable movie.

And the cast should be no problem: simply find a species that has, perhaps inadvertently, perhaps heedlessly, for better or worse, unavoidably and irreversibly, taken over the reins of evolution. All one can hope in addition is that it won't also be necessary to invent a happy ending.

THE BIG BESTSELLERS
ARE AVON BOOKS